Host-Directed Therapies for Tuberculosis

Host-Directed Therapies for Tuberculosis

Editor

Vishwanath Venketaraman

MDPI • Basel • Beijing • Wuhan • Barcelona • Belgrade • Manchester • Tokyo • Cluj • Tianjin

Editor
Vishwanath Venketaraman
Microbiology/Immunology,
Department of Basic Medical Sciences,
College of Osteopathic Medicine of the Pacific,
Western University of Health Sciences
USA

Editorial Office
MDPI
St. Alban-Anlage 66
4052 Basel, Switzerland

This is a reprint of articles from the Special Issue published online in the open access journal *Journal of Clinical Medicine* (ISSN 2077-0383) (available at: https://www.mdpi.com/journal/jcm/special_issues/Tuberculosis_Therapies).

For citation purposes, cite each article independently as indicated on the article page online and as indicated below:

LastName, A.A.; LastName, B.B.; LastName, C.C. Article Title. *Journal Name* **Year**, *Article Number*, Page Range.

ISBN 978-3-03943-501-2 (Hbk)
ISBN 978-3-03943-502-9 (PDF)

Contents

About the Editor . . . vii

Preface to "Host-Directed Therapies for Tuberculosis" . . . ix

Rachel Abrahem, Ruoqiong Cao, Brittanie Robinson, Shalok Munjal, Thomas Cao, Kimberly To, David Ashley, Joshua Hernandez, Timothy Nguyen, Garrett Teskey and Vishwanath Venketaraman
Elucidating the Efficacy of the Bacille Calmette–Guérin Vaccination in Conjunction with First Line Antibiotics and Liposomal Glutathione
Reprinted from: *J. Clin. Med.* **2019**, *8*, 1556, doi:10.3390/jcm8101556 . . . 1

Afsal Kolloli, Pooja Singh, G. Marcela Rodriguez and Selvakumar Subbian
Effect of Iron Supplementation on the Outcome of Non-Progressive Pulmonary *Mycobacterium tuberculosis* Infection
Reprinted from: *J. Clin. Med.* **2019**, *8*, 1155, doi:10.3390/jcm8081155 . . . 27

Meng-Rui Lee, Ming-Chia Lee, Chia-Hao Chang, Chia-Jung Liu, Lih-Yu Chang, Jun-Fu Zhang, Jann-Yuan Wang and Chih-Hsin Lee
Use of Antiplatelet Agents and Survival of Tuberculosis Patients: A Population-Based Cohort Study
Reprinted from: *J. Clin. Med.* **2019**, *8*, 923, doi:10.3390/jcm8070923 . . . 47

Chin-Chung Shu, Shih-Chieh Chang, Yi-Chun Lai, Cheng-Yu Chang, Yu-Feng Wei and Chung-Yu Chen
Factors for the Early Revision of Misdiagnosed Tuberculosis to Lung Cancer: A Multicenter Study in A Tuberculosis-Prevalent Area
Reprinted from: *J. Clin. Med.* **2019**, *8*, 700, doi:10.3390/jcm8050700 . . . 61

Steve Ferlita, Aram Yegiazaryan, Navid Noori, Gagandeep Lal, Timothy Nguyen, Kimberly To and Vishwanath Venketaraman
Type 2 Diabetes Mellitus and Altered Immune System Leading to Susceptibility to Pathogens, Especially *Mycobacterium tuberculosis*
Reprinted from: *J. Clin. Med.* **2019**, *8*, 2219, doi:10.3390/jcm8122219 . . . 71

Yash Dara, Doron Volcani, Kush Shah, Kevin Shin and Vishwanath Venketaraman
Potentials of Host-Directed Therapies in Tuberculosis Management
Reprinted from: *J. Clin. Med.* **2019**, *8*, 1166, doi:10.3390/jcm8081166 . . . 83

Stephen Cerni, Dylan Shafer, Kimberly To and Vishwanath Venketaraman
Investigating the Role of Everolimus in mTOR Inhibition and Autophagy Promotion as a Potential Host-Directed Therapeutic Target in *Mycobacterium tuberculosis* Infection
Reprinted from: *J. Clin. Med.* **2019**, *8*, 232, doi:10.3390/jcm8020232 . . . 95

About the Editor

Vishwanath Venketaraman is a tenured Full Professor with an active research program on tuberculosis. Dr. Venketaraman has published 78 papers and has edited numerous textbooks. He teaches immunology, microbiology, and infectious disease topics to first- and second-year medical students as well as Master students. Dr. Venketaraman, recognized for his teaching and scholarly activities, has received several awards (sixteen) including top honors such as the Distinguished Teacher Award (2017) and the Distinguished Scholar Award (2019) from College of Osteopathic Medicine of the Pacific and from the Western University of Health Sciences. Dr. Venketaraman's research has been continuously funded since 2003. He is currently funded by the NIIH and industry (Your Energy Systems). His research interests include understanding host immune responses against Mycobacterium tuberculosis infection in individuals with HIV and people with type 2 diabetes. His long-term goal is to discover host-directed therapies for tuberculosis.

Preface to "Host-Directed Therapies for Tuberculosis"

TB is considered one of the oldest documented infectious diseases in the world and is believed to be the leading cause of mortality due to a single infectious agent. *Mtb*, the causative agent responsible for TB, continues to afflict millions of people worldwide. Furthermore, one-third of the entire world's population has latent TB. Consequently, there has been a worldwide effort to eradicate and limit the spread of *Mtb* through the use of antibiotics. However, management of TB is becoming more challenging with the emergence of drug-resistant and multi-drug resistant strains of *Mtb*. Furthermore, when administered, many of the anti-TB drugs commonly present severe complications and side effects. Novel approaches to enhance the host immune responses to completely eradicate *Mtb* infection are urgently needed. This Special Issue will, therefore, cover recent advances in the area of host-directed therapies for TB.

Vishwanath Venketaraman
Editor

Journal of
Clinical Medicine

MDPI

Article

Elucidating the Efficacy of the Bacille Calmette–Guérin Vaccination in Conjunction with First Line Antibiotics and Liposomal Glutathione

Rachel Abrahem [1,2], Ruoqiong Cao [3], Brittanie Robinson [2], Shalok Munjal [2], Thomas Cho [2], Kimberly To [1], David Ashley [1,2], Joshua Hernandez [1,2], Timothy Nguyen [2], Garrett Teskey [2] and Vishwanath Venketaraman [1,3,*]

1 Graduate College of Biomedical Sciences, Western University of Health Sciences, Pomona, CA 91766-1854, USA; rachel.abrahem@westernu.edu (R.A.); kimberly.to@westernu.edu (K.T.); david.ashley@westernu.edu (D.A.); joshua.hernandez@westernu.edu (J.H.)
2 College of Osteopathic Medicine of the Pacific, Western University of Health Sciences, Pomona, CA 91766-1854, USA; brittanie.robinson@westernu.edu (B.R.); shalok.munjal@westernu.edu (S.M.); thomas.cho@westernu.edu (T.C.); timothy.nguyen@westernu.edu (T.N.); gteskey@westernu.edu (G.T.)
3 Department of Basic Medical Sciences, College of Osteopathic Medicine of the Pacific, Western University of Health Sciences, Pomona, CA 91766-1854, USA; rcao@westernu.edu
* Correspondence: vvenketaraman@westernu.edu

Received: 18 July 2019; Accepted: 19 September 2019; Published: 27 September 2019

Abstract: *Mycobacterium tuberculosis* (*M. tb*) is the etiological agent that is responsible for causing tuberculosis (TB). Although every year *M. tb* infection affects millions of people worldwide, the only vaccine that is currently available is the Bacille Calmette–Guérin (BCG) vaccine. However, the BCG vaccine has varying efficacy. Additionally, the first line antibiotics administered to patients with active TB often cause severe complications and side effects. To improve upon the host response mechanism in containing *M. tb* infection, our lab has previously shown that the addition of the biological antioxidant glutathione (GSH) has profound antimycobacterial effects. The aim of this study is to understand the additive effects of BCG vaccination and *ex-vivo* GSH enhancement in improving the immune responses against *M. tb* in both groups; specifically, their ability to mount an effective immune response against *M. tb* infection, maintain $CD4^+$ and $CD8^+$ T cells in the granulomas, their response to liposomal glutathione (L-GSH), with varying suboptimal levels of the first line antibiotics isoniazid (INH) and pyrazinamide (PZA), the expressions of programmed death receptor 1 (PD-1), and their ability to induce autophagy. Our results revealed that BCG vaccination, along with GSH enhancement, can prevent the loss of $CD4^+$ and $CD8^+$ T cells in the granulomas and improve the control of *M. tb* infection by decreasing the expressions of PD-1 and increasing autophagy and production of the cytokines interferon gamma IFN-γ and tumor necrosis factor-α (TNF-α).

Keywords: *M. tb*; BCG vaccination; immune exhaustion; glutathione; cytokines; granulomas

1. Introduction

Tuberculosis (TB), caused by *Mycobacterium tuberculosis (M. tb)*, continues to afflict millions of people worldwide. In 2017, approximately 10 million people suffered from active TB and 1.6 million died from this disease [1]. Additionally, one third of the world's population is latently infected with *M. tb*. Individuals infected with *M. tb* have a 5–15% lifetime risk of developing an active disease; however, immunocompromised patients, such as people living with diabetes, malnutrition, human immunodeficiency virus (HIV), or those who use tobacco, have a higher risk of developing active TB [1]. Common symptoms of active pulmonary TB are coughs with a bloody sputum, chest pains, weakness, weight loss, fever, and night sweats, eventually resulting in death when untreated [1]. *M. tb* infection occurs due to inhalation of infectious aerosolized droplets, and the bacteria become seeded in the lower respiratory tract where there is an enrichment of alveolar macrophages.

M. tb infection is initiated when the inhaled organisms are phagocytosed by these alveolar macrophages [2]. In an immune-competent individual, the immune system is able to mount a formidable response against *M. tb*, resulting in the formation of a solid and robust granuloma. Composed of a compact aggregate of immune cells [3]. Mature macrophages in the granuloma can fuse into multinucleated giant cells or differentiate into foam cells and epithelioid cells [3]. Alongside macrophages, other cells, such as neutrophils, dendritic cells, natural killer cells, fibroblasts, $CD4^+$ T cells, and cytotoxic $CD8^+$ T cells, are also recruited into the granuloma via cytokine mediation, leading to containment of the *M. tb* infection [3]. The effector responses inside the granulomas along with a lack of nutrients and oxygen causes *M. tb* to become dormant and remain latent in a nonreplicating state. The contained *M. tb* within a granuloma in the lungs is commonly referred to as latent tuberculosis (LTBI). A breakdown of immune responses designed to contain the infection in immunocompromised individuals can result in reactivation of *M. tb* [4]. This dysregulation promotes liquification of caseum and replication of *M. tb*, thereby promoting cavity formation and the release of *M. tb* to the exterior during coughing, ultimately spreading the infection to other parts of the lungs [5]. Active *M. tb* is able to deflect many host defense mechanisms via the cord factor, preventing phagosome-lysosome fusion and the degradation of the bacilli [6].

In order to effectively contain *M. tb* within the granuloma, proper cytokine-mediated signaling is essential to promote the necessary aggregation of cells [7]. Cytokines, such as interferon gamma (IFN-γ) and tumor necrosis factor-α (TNF-α), play a critical role in both the innate and adaptive immune responses against *M. tb* infection [8]. TNF-α produced by macrophages induces the formation and maintenance of the granuloma. The T-helper 1 (Th1) subset of $CD4^+$ T cells releases IFN-γ to activate effector mechanisms in macrophages to not only kill *M. tb* intracellularly but also enhance the effector functions of natural killer cells and cytotoxic T lymphocytes ($CD8^+$) T-cells [9].

Once activated, $CD8^+$ T cells and natural killer cells will then produce antimicrobial peptides, perforin and granulysin, to destroy intracellular *M. tb* and the host cells.

The programmed death receptor 1 (PD-1), a negative regulator of activated T cells, is markedly upregulated on the surface of pathogen-specific $CD8^+$ T cells in mice [10]. Blockage of this pathway restores the $CD8^+$ T cell function and reduces the microbial load [10]. PD-1 is also expressed on the surface of $CD4^+$ T cells, with a positive correlation in regard to the microbial load and an inverse correlation with the $CD4^+$ T cell count.

Although the immune system has a robust defense system in place to combat *M. tb* infection, it cannot always contain the infection. For this reason, host directed therapy is often required.

The Center for Disease Control (CDC) recommends four anti-TB agents to form the core of the treatment regimen for patients with active TB [11]. These drugs include isoniazid (INH), rifampicin (RIF), ethambutol (EMB), and pyrazinamide (PZA). This extensive drug regimen often leads to noncompliance to the TB treatment, leading to multidrug-resistant (MDR)-TB, which is typically resistant to both INH and RIF [12].

Although a myriad of antibiotics can be used to treat this mycobacterial infection, there is only one vaccine available to prime the immune system against *M. tb*, and that is the *Mycobacterium bovis* bacille Calmette–Guérin (BCG) vaccine [13]. BCG used in this vaccine is an attenuated strain of *M. bovis* [13]. The WHO recommends that infants in countries with a high risk of *M. tb* infection be immunized with the BCG vaccine soon after birth [13]. Because the incidence of TB is low in the United States, it is not recommended for infants to be administered this vaccination. Additionally, estimates of the protective efficacy of the BCG vaccine against adult pulmonary TB very widely, ranging from 0 to 80% [13].

Glutathione (GSH), a tripeptide antioxidant composed of glutamine, cysteine, and glycine, is found ubiquitously amongst all cell types. GSH prevents cellular damage by detoxifying reactive oxygen species (ROS) [14]. GSH exists in both a reduced state (rGSH) and an oxidized form (GSSG) [14]. rGSH contains the antioxidant properties, while GSSG is a simple byproduct of the oxidation of GSH and has no antioxidant effects [14]. When exposed to ROS, two molecules of rGSH are converted to GSSG and water [14] Mycobacteria possess an alternative thiol, mycothiol, rather than GSH, to regulate their redox homeostasis [15]. Due to this property, the presence of millimolar concentrations of GSH (physiological concentrations) inside infected macrophages can lead to inhibition in the growth of *M. tb* [15]. Our laboratory has previously demonstrated that GSH-enhancement by N-acetyl cysteine (NAC) supplementation resulted in a significant reduction of *M. tb* burden among both healthy and diabetic individuals [15] Additionally, enhancement of GSH by means of NAC has the potential implications of not only reducing the toxicity of anti-TB medications through GSH's redox potential but may possibly permit lower antibiotic dosage to promote enhanced patient compliance [15]. For this reason, our lab tested the effects of liposomal glutathione (L-GSH) in the presence and absence of sub-optimal concentration of INH and PZA in improving the ability of immune cells isolated from BCG-vaccinated and non-vaccinated individuals to control *M. tb* infection.

In this study, we determined the additive effects of BCG vaccination and *in vitro* GSH-enhancement in improving the ability of immune cells to control *M. tb* infection by measuring the differences in the immune responses between vaccinated and non-vaccinated individuals. Fluorescent staining and other antibody assays were also performed to determine the underlying mechanistic differences in the ability of immune cells from vaccinated and unvaccinated groups to respond to L-GSH, cytokine production, the surface expression of CD4, CD8, PD-1, and their ability to induce autophagy.

2. Materials and Methods

2.1. Peripheral Blood Mononuclear Cell Isolation

Peripheral blood mononuclear cells (PBMCs) were isolated from the whole blood of both BCG-vaccinated and non-BCG-vaccinated participants. Whole blood was layered in a 1:1 ratio onto ficoll histopaque (Sigma, St. Louis, MO, USA), a high density-pH neutral polysaccharide solution, for density gradient centrifugation (1800 rpm for 30 min) [15]. The PBMCs at the interface were aspirated, washed twice with sterile 1X PBS (Sigma, St. Louis, MO, USA), and resuspended in Roswell Park Memorial Institute (RPMI) (Sigma, St Louis, MO, USA) with 5% human AB serum (Sigma, St. Louis, MO, USA). PBMC counts were determined by trypan blue exclusion staining.

2.2. Generation of In Vitro Granulomas

Our laboratory had successfully established an *in vitro* human granuloma model using PBMCs, isolated from healthy subjects and individuals with type 2 diabetes [15–18]. These granulomas [15–18] exhibit a physically well-demarcated aggregation of mononuclear cells with a denser central core descending towards the periphery, which can be seen in the in vitro granulomas. Multi-nucleated giant cells (MNGs hallmark of granulomas), T cells, and activated macrophages were also seen. These features are reminiscent of early stage, cellular lung granulomas in experimental animal models of TB, including rabbits and non-human primates. Such granulomas are also noted in the lungs of mice during chronic *M. tb* infection. Using our previously published protocol, we developed in vitro granulomas for the current study. Isolated PBMCs from the two study groups resuspended in RPMI were infected with the Erdman strain of *M. tb* at a multiplicity of infection (MOI) of 0.1:1 cell ratio. 500 µL of the cell suspension containing PBMCs and *M. tb* were added to the 24-well plates. To ensure proper adhesion of isolated immune cells, 24-well plates (Corning, Corning, NY, USA) were coated with 0.001% poly-lysine (Sigma, St. Louis, MO, USA) overnight [15,16]. PBMCs (6×10^5 cells/well) were distributed into the poly-lysine coated 24-well plates [15,16]. PBMCs in the wells were either sham-treated or treated with the minimum inhibitory concentration (MIC), a 1.10 dilution, and a 1.100 dilution of two first-line antibiotics with and without 120 µM of liposomal glutathione (L-GSH (Your Energy Systems)). This comprised INH (0.125 micrograms/mL) standalone, 120 µM of L-GSH, 1/10 INH (0.0125 micrograms/mL) standalone, 120 µM of L-GSH, 1/100 INH (0.00125 micrograms/mL) standalone, 120 µM of L-GSH, PZA (50 micrograms/mL), 120 µM of L-GSH, 1/10 PZA

(5 micrograms/mL) standalone, or 120 μM of L-GSH and 1/100 PZA (0.5 micrograms/mL) standalone or 120 μM of L-GSH. All tissue culture plates with infected PBMCs were maintained at 37 °C with 5% CO_2 until they were terminated at 8 days post infection.

2.3. Termination of Granulomas

Following 8 days post infection, the minimum time needed for granuloma formation, *in vitro* granulomas were terminated to determine the intracellular survival of *M. tb* [15,16]. To terminate, the supernatants of each category were aspirated and collected into eppendorf tubes separated by a treatment group, and 250 μL of ice cold, sterile 1× PBS was replaced in lieu of the supernatants followed by gentle scraping of the wells [15,16]. Scraping was done to ensure maximum recovery of granuloma lysates from the wells.

2.4. Colony Forming Units

Collected supernatants and lysates from termination were plated on 7H11 agar media (Hi Media, Santa Maria, CA, USA) enriched with Albumin Dextrose Complex (ADC) (GEMINI, Calabasas, CA, US). They were incubated for a minimum of 3 weeks and 3 days to evaluate the mycobacterial survival under the different treatment conditions by counting the colony forming units (CFUs).

2.5. Cytokine Measurements

To measure cytokine levels, the sandwich enzyme-linked immunosorbent assay (ELISA) technique was used. The assay was performed via the manufacturer's protocol (Invitrogen, Carlsbad, CA, USA). The cytokines measured were IFN-γ and TNF-α in the supernatants at 8 days post-infection to determine the effects of the antibiotics with and without L-GSH treatments on cytokine levels in BCG-vaccinated and non-vaccinated individuals.

2.6. Glutathione Measurements

Levels of GSH from the granulomas of non-vaccinated and BCG-vaccinated subjects were measured by the colorimetric method using an assay kit from Arbor Assay (K006-H1). Granuloma lysates were mixed in a 1:1 ratio with cold 5% sulfosalicylic acid (SSA), incubated for 10 min at 4 °C, followed by centrifugation at 14,000 rpm for 10 min. The GSH was measured in the lysates following the manufacturer's instructions. The reduced GSH (rGSH) was calculated by subtracting the oxidized glutathione (GSSG) from the total GSH.

2.7. Staining and Imaging Techniques

Each trial contained designated wells for fluorescent and light microscopic studies. Cover glasses were allotted into 24-well plates for granuloma formation observation. The cover glasses were fixed with 4% paraformaldehyde (PFA) for 1 h at room temperature and washed three times with 1× PBS for 5 min to remove cell debris. Fixed granulomas were then stained with Hematoxylin and Eosin (H&E) (Poly Scientific, Bay Shore, NY, USA) for 2 min at room temperature and destained with deionized water. The granuloma-stained cover glasses were mounted onto glass slides with HistoChoice mounting media. Fixed granulomas on cover glasses were also permeabilized with Triton X for 2 min and stained overnight with antibodies conjugated with fluorescent markers (CD4-PE, CD8-PE, and LC3B-PE). Cover glasses were washed with phosphate buffer saline (PBS) and mounted on clean glass slides with mounting media containing 4',6-diamidino-2-phenylindole DAPI For PD1 staining, fixed granulomas on cover glasses were permeabilized with Triton X for 2 min and incubated overnight with anti-PD1 (Pro-Sci), followed by incubation for another 2 h with c-Myc. Cover glasses were then incubated overnight with secondary antibodies (mouse anti-human) and conjugated with fluorescein isothiocyanate (FITC). Cover glasses were mounted using a mounting media containing DAPI. Slides were observed under the fluorescent microscope. Fluorescent images were captured, and the fluorescent intensity was quantified using the ImageJ software (version 8, GraphPad, San Diego, CA, USA).

2.8. Statistical Analysis

Statistical data analysis was performed using GraphPad Prism Software 8 using the unpaired t-test with Welch correction for two sampled graphs. A one-way ANOVA (analysis of variance) was performed for samples with greater than two categories with Tukey corrections. Reported values are the means with each respective category. A $p < 0.05$ was considered significant. The p value style consisted of 0.1234 as not significant, 0.0332 with one asterisk (*), 0.0021 with two asterisks (**), 0.0002 with three asterisks (***), and less than 0.0001 with four asterisks (****). A hash mark (#) indicates categories compared to control, and an asterisk indicates categories compared to the previous category directly before it.

3. Results

3.1. Survival of the Erdman Strain of *M. tb* in the In Vitro Granulomas

We first tested the effects of L-GSH in controlling the growth of *M. tb* inside in vitro granulomas derived from non-vaccinated and BCG-vaccinated subjects. Approximately, 25 μL of granuloma lysates were plated on 7H11 growth media and incubated for four weeks to allow adequate time for *M. tb* growth. In both non-vaccinated and BCG-vaccinated individuals, there was a significant reduction in the bacterial load when standalone L-GSH was added (Figure 1A,B). We then measured the effects of PZA added at various concentrations in the presence and absence of L-GSH in altering the viability of *M. tb*. In the non-vaccinated group, there was a significant reduction in the viability of *M. tb* when PZA was added at MIC and at the 1/10 lower dilution in the presence and absence of L-GSH when compared to the untreated control (Figure 1C). In the BCG-vaccinated group, there was a significant reduction in the viability of *M. tb* with the addition of L-GSH at all tested concentrations of PZA when compared to the untreated control category and to the PZA-alone treated groups (MIC, 10 and 100 times lower than MIC concentrations). Notably, in the BCG-vaccinated group, PZA + L-GSH 120 resulted in complete clearance of *M. tb* (Figure 1D). Hematoxylin and Eosin staining was performed to observe the morphology of the granuloma-like structures. Not surprisingly, we witnessed more solid and robust granulomas when the bacterial load was higher. Correspondingly, the aggregation of the granuloma was not as dense when the infection was cleared (Figure 1A,B). Additionally, we compiled the results from the *M. tb* survival assays from the non-vaccinated and BCG-vaccinated individuals to compare the same treatment categories. The BCG-vaccinated subjects displayed a significantly greater ability to kill *M. tb* and/or containment potential than the non-vaccinated control. L-GSH treatment resulted in a similar trend, exhibiting improved killing of *M. tb* in the granulomas from BCG-vaccinated individuals when compared to the non-vaccinated subjects (Figure 1E). In regard to the comparison of the PZA treated granulomas from non-vaccinated and BCG-vaccinated subjects, there was a significant reduction in the viability of *M. tb* in all categories of the BCG-vaccinated individuals when compared to the non-vaccinated individuals of the same treatment category (Figure 1F).

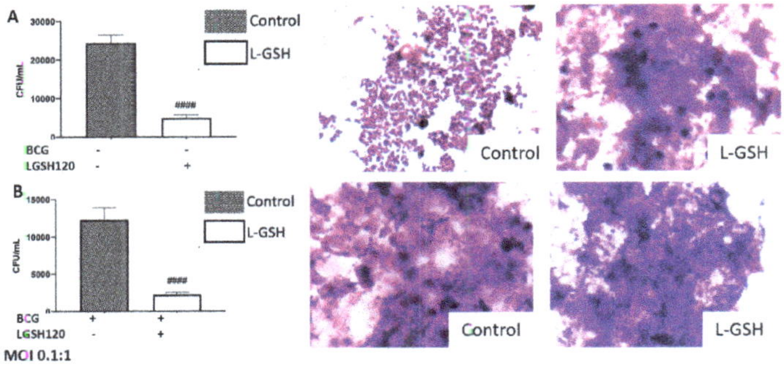

Figure 1. *Cont.*

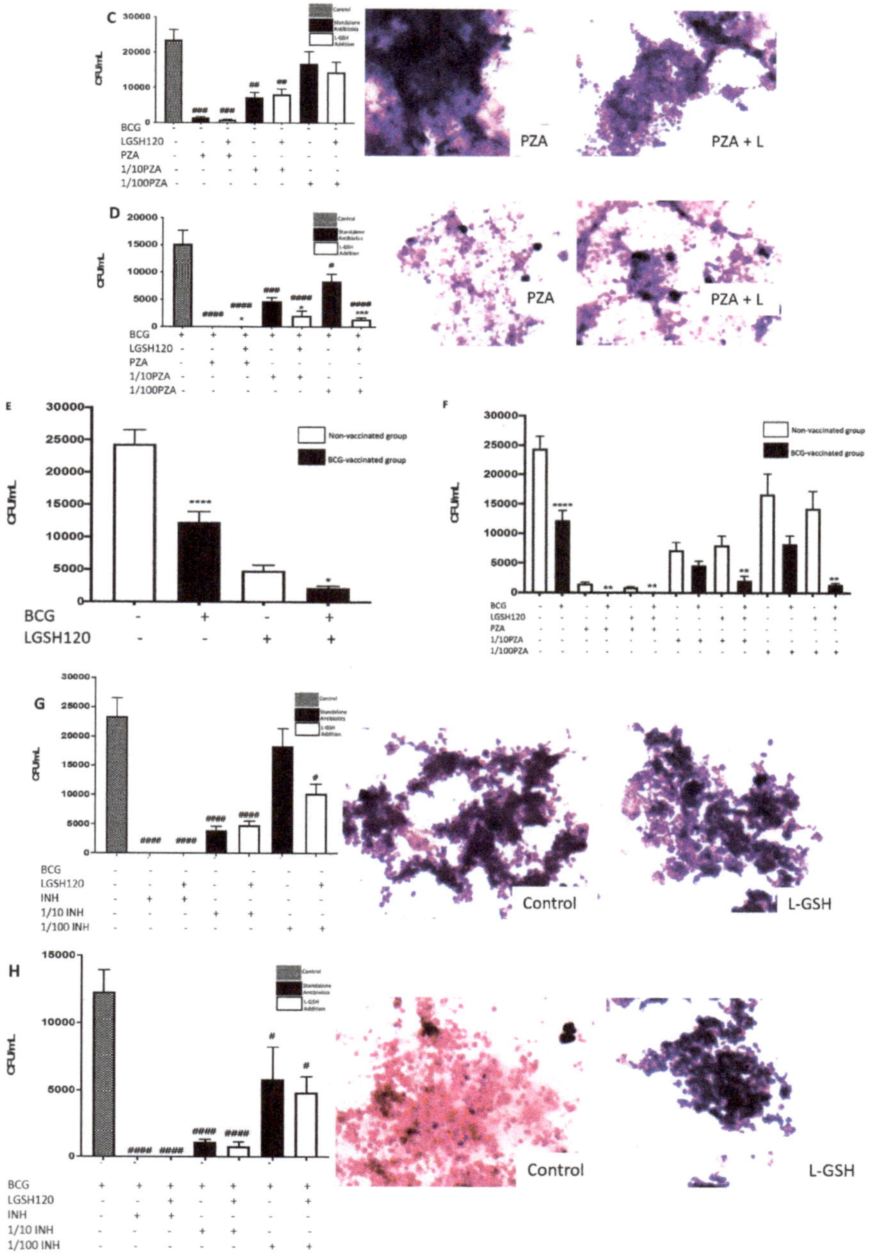

Figure 1. *Cont.*

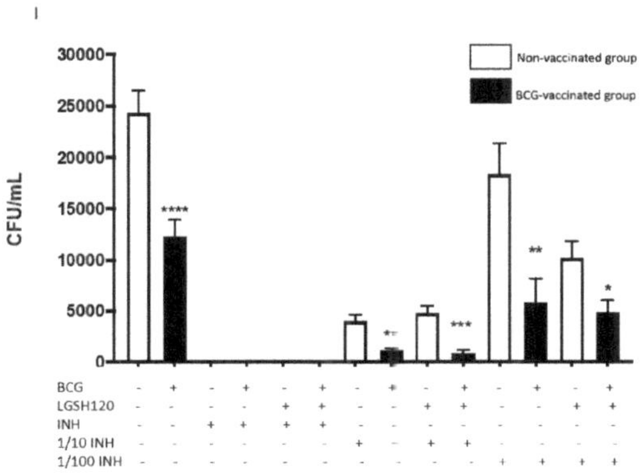

Figure 1. Survival of the Erdman strain of *Mycobacterium tuberculosis* treated with Pyrazinamide and Isoniazid in media. (**A**) *M. tb* growth in 7H11 agar media from non-vaccinated subjects with liposomal glutathione (L-GSH) treatment along with Hematoxylin and Eosin (H&E) images of granuloma like structures from eight-day terminated samples; (**B**) *M. tb* growth in 7H11 agar media from Bacille Calmette–Guérin (BCG) vaccinated subjects with L-GSH treatment along with H&E images of granuloma-like structures from eight-day terminated samples; (**C**) *M. tb* growth in 7H11 agar media with pyrazinamide (PZA), 1/10 PZA, and 1/100 PZA treatment in the presence or absence of L-GSH in non-vaccinated subjects; (**D**) *M. tb* growth in 7H11 agar media with PZA, 1/10 PZA, and 1/100 PZA treatment in the presence or absence of L-GSH in BCG-vaccinated subjects. (**E**) *M. tb* growth in 7H11 agar media from non-vaccinated (white bar) and BCG-vaccinated (black bar) groups with sham treatment and L-GSH treatment; (**F**) *M. tb* growth in 7H11 agar media in non-vaccinated and BCG-vaccinated groups with PZA treatment. Data represents ± SE(standard error) rom experiments performed from 14 different subjects. An unpaired t-test with Welch corrections was used in Figure (**A**). Analysis of Figures B–F utilized a one-way ANOVA (analysis of variance) with Tukey test. The *p* value style consisted of 0.1234 as not significant, 0.0332 with one asterisk (*), 0.0021 with two asterisks (**), 0.0002 with three asterisks (***), and less than 0.0001 with four asterisks (****). A hash mark (#) indicates categories compared to control, and an asterisk indicates categories compared to the previous category directly before it. (**G**) *M. tb* growth in 7H11 agar media with isoniazid (INH), 1/10 INH, and 1/100 INH treatment in the presence or absence of L-GSH in non-vaccinated subject; (**H**) *M. tb* growth in 7H11 agar media with INH, 1/10 INH, and 1/100 INH treatment in the presence or absence of L-GSH in BCG-vaccinated subjects; (**I**) *M. tb* growth in 7H11 agar media in non-vaccinated and BCG-vaccinated groups with INH treatment. Data represents ± SE from experiments performed from 14 different subjects. Analysis of figures utilized a one-way ANOVA with Tukey test. The *p* value style consisted of 0.1234 as not significant, 0.0332 with one asterisk (*), 0.0021 with two asterisks (**), 0.0002 with three asterisks (***), and less than 0.0001 with four asterisks (****). A hash mark (#) indicates categories compared to control, and an asterisk indicates categories compared to the previous category directly before it.

The effects of INH in altering the viability of *M. tb* in the granulomas of non-vaccinated and BCG-vaccinated subjects was also tested. In the non-vaccinated group, there was a significant reduction in the viability of *M. tb* (almost undetectable colonies) when INH was added at MIC. There was a significant reduction in CFUs in all categories, excluding the 1/100 INH category (Figure 1G). In the BCG-vaccinated group, there was a significant reduction in the viability of *M. tb* at MIC and in all categories (Figure 1H). When comparing INH treated granulomas in non-vaccinated and BCG-vaccinated subjects, granulomas from BCG-vaccinated subjects controlled the *M. tb* infection more effectively compared to the non-vaccinated groups in the presence and absence of INH and/or L-GSH (Figure 1I).

3.2. Levels of the Reduced Form of GSH in the In Vitro Treated Granulomas

GSH levels were measured from the cellular lysates of the PBMCs using an assay kit from Arbor Assays. The untreated granulomas from BCG-vaccinated individuals displayed higher levels of the reduced form of GSH than non-vaccinated individuals. L-GSH treatment resulted in increased levels of GSH in both the groups; however, there was a significant increase in the levels of GSH from the granulomas of BCG-vaccinated individuals (Figure 2A). Treatment of granulomas from BCG-vaccinated subjects with PZA in combination with L-GSH resulted in a significant increase in the levels of GSH compared to the non-vaccinated group (Figure 2B). The levels of reduced forms of GSH were also measured in the lysates of the eight-day terminated samples. In granulomas treated with INH, BCG-vaccinated individuals had a significant increase in GSH in all categories in the presence and absence of L-GSH (Figure 2C).

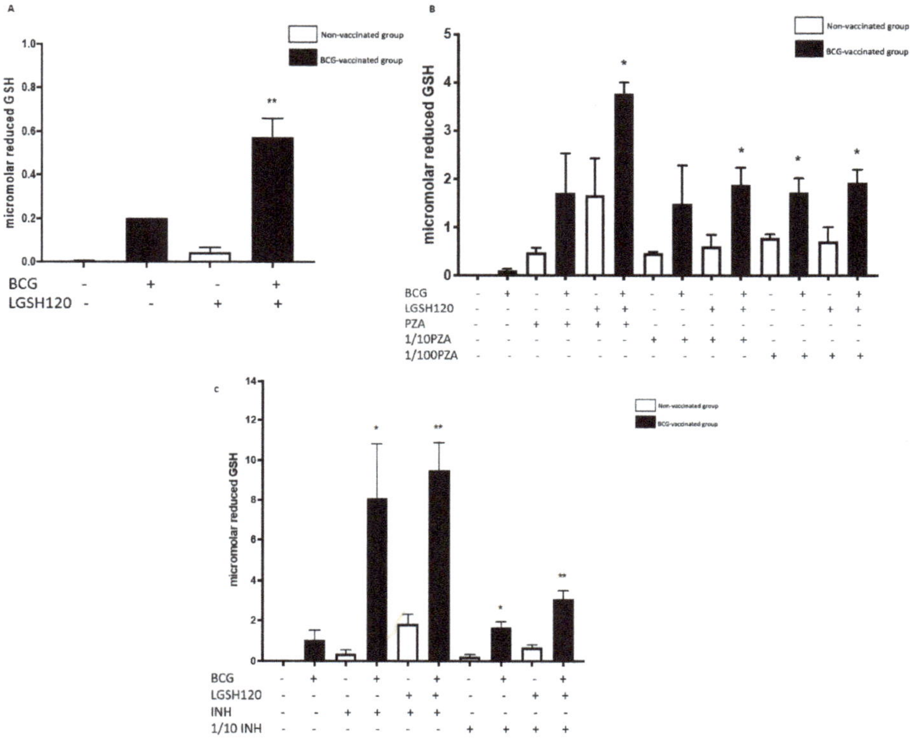

Figure 2. Levels of reduced GSH in PZA and INH treated granulomas. The GSH assay was performed by the colorimetric method using an Arbor Assays kit. (**A**) Comparison of reduced GSH levels in non-vaccinated and BCG-vaccinated groups; (**B**) GSH measurements in PZA treated granulomas from non-vaccinated and BCG-vaccinated subjects. Data represents ±SE from experiments performed from 14 different subjects. The p value style consisted of 0.1234 as not significant, 0.0332 with one asterisk (*), 0.0021 with two asterisks (**), 0.0002 with three asterisks (***), and less than 0.0001 with four asterisks (****). (**C**) The GSH assay was performed by the colorimetric method using an Arbor Assays kit. GSH measurements in INH treated granulomas from non-vaccinated and BCG-vaccinated subjects. Data represents ± SE from experiments performed from 14 different subjects. Analysis of figures utilized a one-way ANOVA with Tukey test. The p value style consisted of 0.1234 as not significant, 0.0332 with one asterisk (*), 0.0021 with two asterisks (**), 0.0002 with three asterisks (***), and less than 0.0001 with four asterisks (****).

3.3. Expression of CD4 in the In Vitro Granulomas

CD4 T cell expression was measured with an anti-CD4 antibody, conjugated with Phycoerythrin-Cy5 (PE-Cy5). Analysis of CD4 T cells were corrected with the average mean fluorescence of 4′,6-diamidino-2-phenylindole (DAPI). The addition of standalone L-GSH resulted in a significant increase of the CD4 mean fluorescence in both vaccinated and non-vaccinated groups (Figure 3A,B). CD4 expression levels were then measured from the pyrazinamide treatment categories in the presence and absence of L-GSH, both in the non-vaccinated group and the BCG-vaccinated group. In the non-vaccinated group, there was a significant increase in CD4 expression with L-GSH addition when compared to both the control group and the standalone PZA. A dilution of 1/100 PZA standalone treatment resulted in a significant increase in CD4 expression when compared to the control (Figure 3C). In the BCG-vaccinated group, L-GSH addition resulted in a significant increase in CD4 expression when compared to the sham treated group. The categories of standalone MIC PZA and 1/10 PZA resulted in a significant increase in CD4 expression when compared to the control. Additionally, there was a significant increase in the expression of CD4 in PZA + L-GSH and 1/10PZA + L-GSH categories when compared to their standalone counterparts (Figure 3D). The compiled graph comparing the sham treated versus L-GSH treated granulomas from non-vaccinated and BCG-vaccinated subjects demonstrated a significant increase in the mean fluorescent intensity of CD4 in BCG-vaccinated subjects with the addition of L-GSH (Figure 3E). In comparison to the non-vaccinated group, PZA treatment resulted in a significant increase in CD4 expression in the granulomas from BCG-vaccinated groups (Figure 3F). L-GSH + PZA treatment resulted in a further increase in the expression of CD4 in the granulomas from BCG-vaccinated subjects compared to the non-vaccinated group (Figure 3F).

There is an increase in the viability potential of CD4 when L-GSH adjunctive treatment is added. In the compiled data comparing the non-vaccinated group to the BCG-vaccinated group, in INH treated granulomas, there was a significant increase in CD4 expression in the BCG-vaccinated groups in most categories, except for MIC and 1/10 INH + L-GSH (Figure 3G,H).

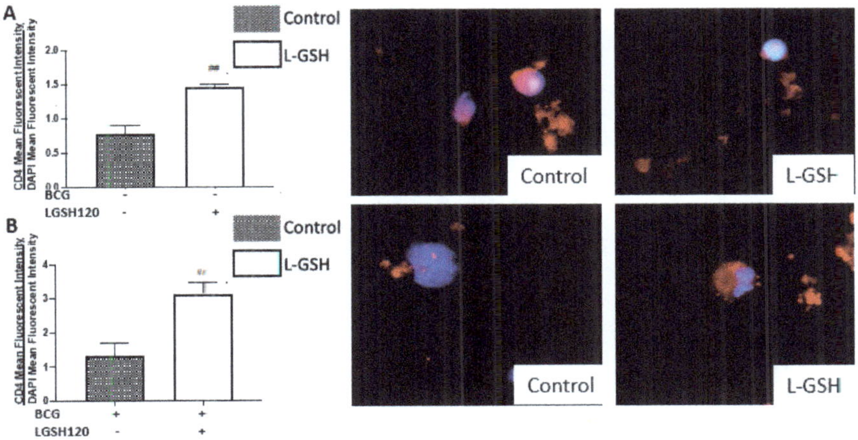

Figure 3. *Cont.*

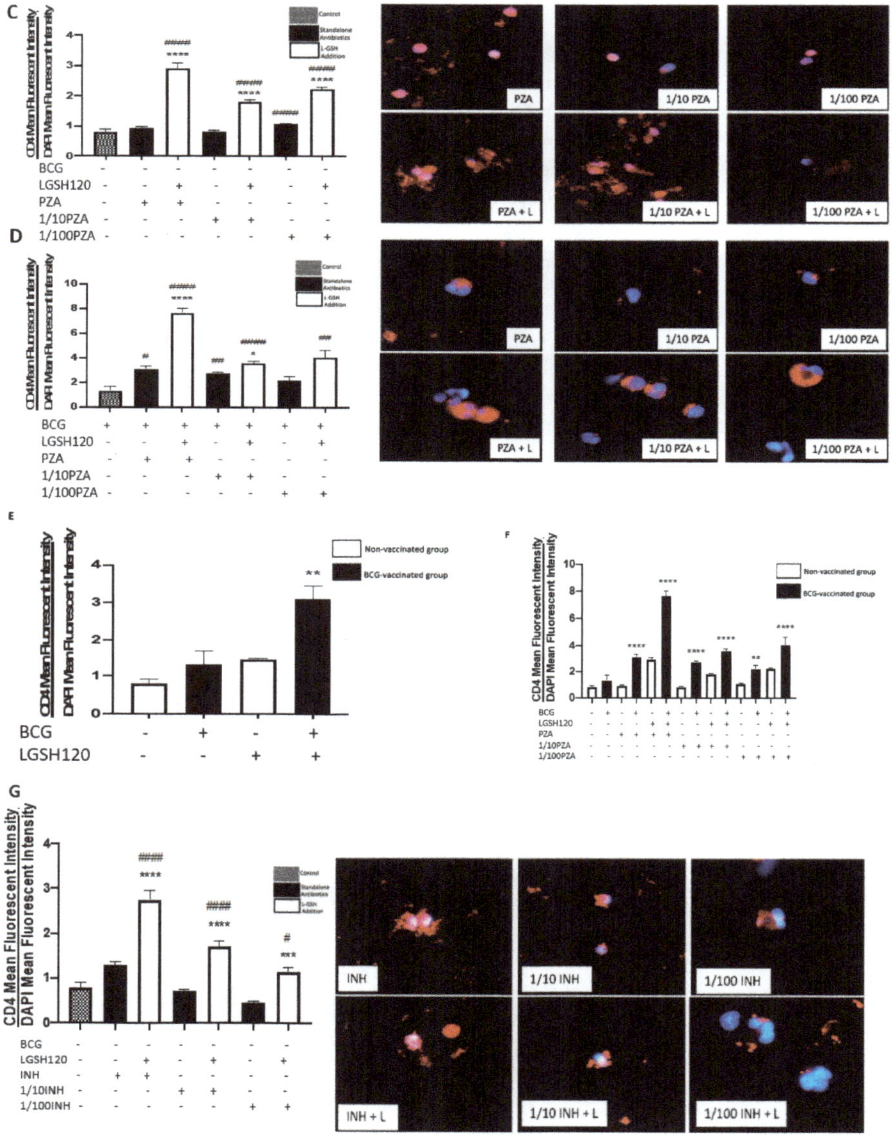

Figure 3. *Cont.*

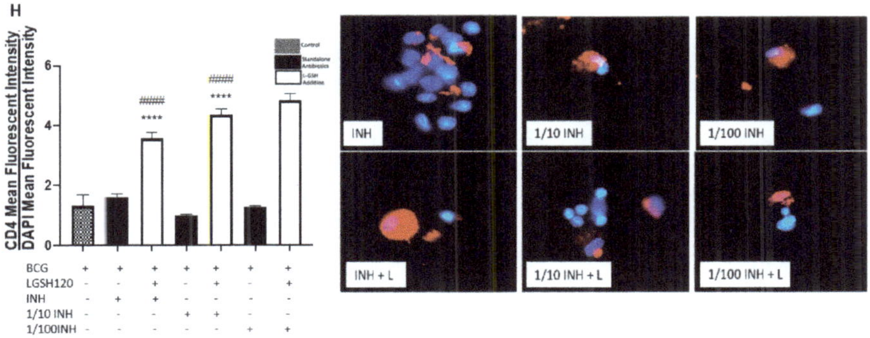

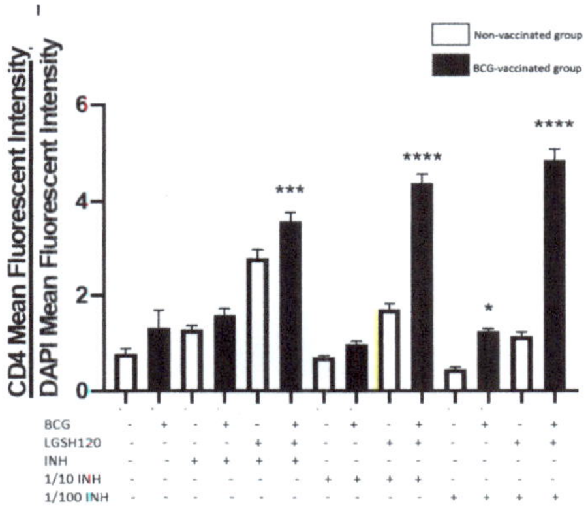

Figure 3. Expression of CD4 in PZA and INH treated granulomas. CD4 T cells were measured with an anti-CD4 antibody conjugated with (Phycoerythrin-Cy5) PE-Cy5. The nuclei of cells were stained with (4′,6-diamidino-2-phenylindole) DAPI. The mean fluorescence of CD4 was corrected with the mean DAPI fluorescence. (**A**) CD4 mean fluorescence from non-vaccinated subjects with L-GSH treatment along with CD4 fluorescent images; (**B**) CD4 mean fluorescence from BCG vaccinated subjects with L-GSH treatment along with CD4 fluorescent images; (**C**) CD4 mean fluorescence with PZA, 1/10 PZA, and 1/100 PZA treatment in the presence or absence of L-GSH in non-vaccinated subjects; (**D**) CD4 mean fluorescence with PZA, 1/10 PZA, and 1/100 PZA treatment in the presence or absence of L-GSH in BCG-vaccinated subjects; (**E**) CD4 mean fluorescence in non-vaccinated and BCG-vaccinated groups with sham treatment and L-GSH treatment; (**F**) CD4 mean fluorescence in non-vaccinated and BCG-vaccinated groups with PZA treatment. Data represents ± SE from experiments performed from 14 different subjects. An unpaired t-test with Welch corrections was used in Figure (**A**). Analysis of Figures B–F utilized a one-way ANOVA with Tukey test. The p value style consisted of 0.1234 as not significant, 0.0332 with one asterisk (*), 0.0021 with two asterisks (**), 0.0002 with three asterisks (***), and less than 0.0001 with four asterisks (****). A hash mark (#) indicates categories compared to control, and an asterisk indicates categories compared to the previous category directly before it (**G**) CD4 mean fluorescence with INH, 1/10 INH, and 1/100 INH treatment in the presence or absence of L-GSH in non-vaccinated subjects; (**H**) CD4 mean fluorescence with INH, 1/10 INH, and 1/100 INH treatment in the presence or absence of L-GSH in BCG-vaccinated subjects; (**I**) CD4 mean fluorescence in non-vaccinated and BCG-vaccinated groups with INH treatment. Data represents ± SE from experiments performed from 14 different subjects. The analysis of figures utilized a one-way ANOVA with Tukey test. The p value style consisted of 0.1234 as not significant, 0.0332 with one asterisk (*), 0.0021 with two asterisks (**), 0.0002 with three asterisks (***), and less than 0.0001 with four asterisks (****). A hash mark (#) indicates categories compared to the control, and an asterisk indicates categories compared to the previous category directly before it.

3.4. Expression of CD8 in the In Vitro Granulomas

CD8 T cell expression was measured with an anti-CD8 antibody, conjugated with PE-Cy5. Analysis of CD8 expression was corrected with the average mean fluorescence of DAPI. Individuals vaccinated with BCG displayed the ability to maintain an elevated number of CD8 T cells in the granulomas compared to the non-vaccinated individuals. The addition of L-GSH significantly increased the already high levels of CD8 expression in the granulomas from BCG-vaccinated subjects (Figure 4A,B). A similar trend was found with the PZA treatment groups. Among the non-vaccinated group, the addition of L-GSH + PZA also significantly increased the levels of CD8 expression when compared to the control group and to its counterpart with standalone antibiotics, with the exception of 1/100 PZA + L-GSH (Figure 4C). However, in the BCG-vaccinated group, there was a significant increase in CD8 expression in all the PZA + L-GSH treatment groups (Figure 4D). This can be explained due to the restorative effects L-GSH exhibits to diminish T cell exhaustion. When directly comparing the non-vaccinated to BCG-vaccinated groups, there was no significance between the two sham-treated and L-GSH-treated categories (Figure 4E). The general trend with PZA treatment demonstrates that, on average, BCG-vaccinated subjects possess higher levels of CD8 expression, with significance found in the 1/10 PZA and 1/100 PZA + L-GSH treatment groups (Figure 4F).

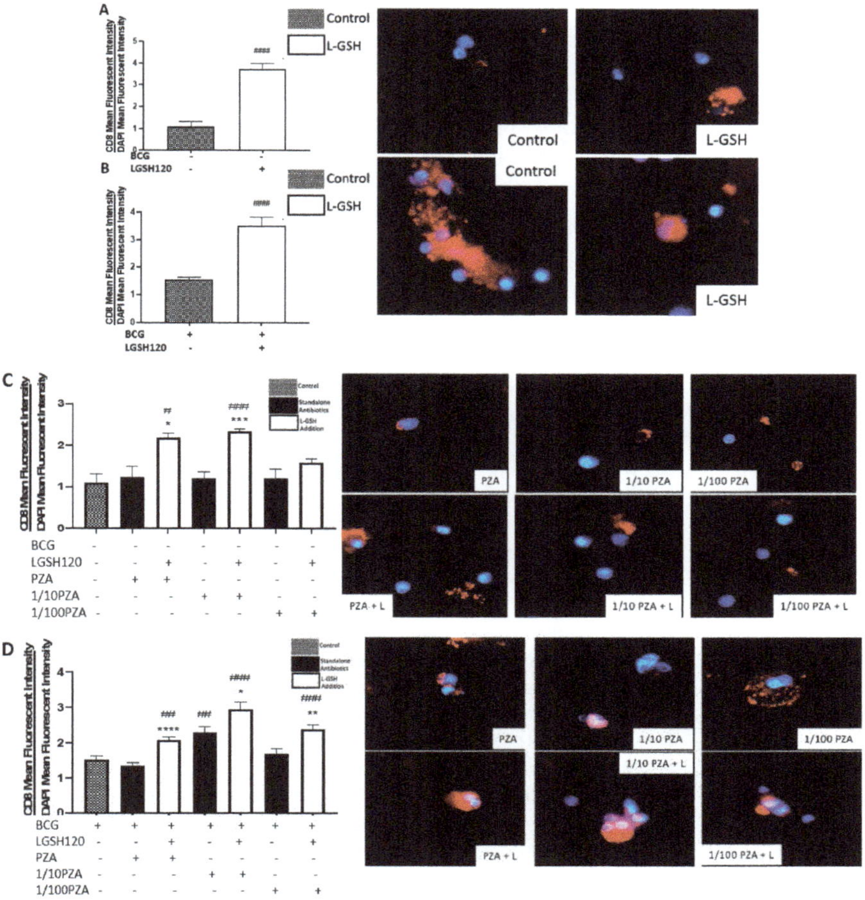

Figure 4. *Cont.*

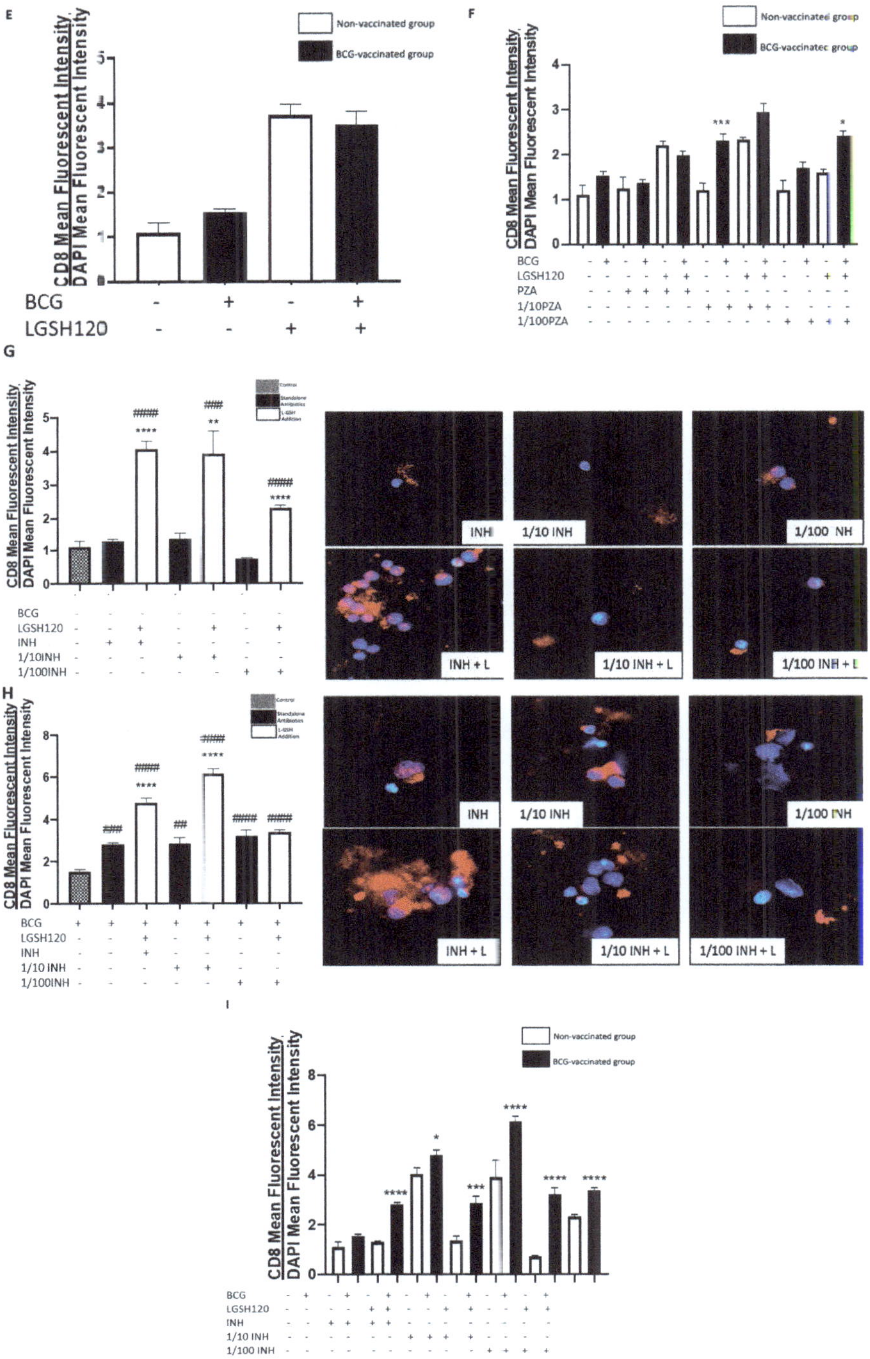

Figure 4. Expression of CD8 in PZA and INH treated granulomas. CD8 T cells were measured with an anti-CD8 antibody conjugated with PE-Cy5. The nuclei of cells were stained with DAPI. The mean

fluorescence of CD8 was corrected with the mean DAPI fluorescence. (**A**) CD8 mean fluorescence from non-vaccinated subjects with L-GSH treatment along with CD4 fluorescent images; (**B**) CD8 mean fluorescence from BCG-vaccinated subjects with L-GSH treatment along with CD8 fluorescent images; (**C**) CD8 mean fluorescence with PZA, 1/10 PZA, and 1/100 PZA treatment in the presence or absence of L-GSH in non-vaccinated subjects; (**D**) CD8 mean fluorescence with PZA, 1/10 PZA, and 1/100 PZA treatment in the presence or absence of L-GSH in BCG-vaccinated subjects; (**E**) CD8 mean fluorescence in non-vaccinated and BCG-vaccinated groups with sham treatment and L-GSH treatment; (**F**) CD8 mean fluorescence in non-vaccinated and BCG-vaccinated groups with PZA treatment. Data represents ± SE from experiments performed from 14 different subjects. An unpaired t-test with Welch corrections was used in Figure (A). Analysis of Figures B–F utilized a one-way ANOVA with Tukey test. The *p* value style consisted of 0.1234 as not significant, 0.0332 with one asterisk (*), 0.0021 with two asterisks (**), 0.0002 with three asterisks (***), and less than 0.0001 with four asterisks (****). A hash mark (#) indicates categories compared to the control, and an asterisk indicates categories compared to the previous category directly before it. (**G**) CD8 mean fluorescence with INH, 1/10 INH, and 1/100 INH treatment in the presence or absence of L-GSH in non-vaccinated subjects; (**H**) CD8 mean fluorescence with INH, 1/10 INH, and 1/100 INH treatment in the presence or absence of L-GSH in BCG-vaccinated subjects; (**I**) CD8 mean fluorescence in non-vaccinated and BCG-vaccinated groups with INH treatment. Data represents ± SE from experiments performed from 14 different subjects. Analysis of figures utilized a one-way ANOVA with Tukey test. The *p* value style consisted of 0.1234 as not significant, 0.0332 with one asterisk (*), 0.0021 with two asterisks (**), 0.0002 with three asterisks (***), and less than 0.0001 with four asterisks (****). A hash mark (#) indicates categories compared to the control, and an asterisk indicates categories compared to the previous category directly before it.

In the presence of L-GSH, there was a significant increase in the levels of CD8 T cells compared to the sham-treated and INH alone treated groups in the non-vaccinated group (Figure 4G). In the BCG-vaccinated group, there was a significant increase in CD8 expression in all categories when compared to the control group. Additionally, in the INH + L-GSH and 1/10 INH + L-GSH treatment groups, there was a significant increase when compared to their standalone antibiotic counterparts (Figure 4H). The mean fluorescence of CD8 in INH-treated granulomas showed a significant increase in BCG-vaccinated subjects compared to non-vaccinated subjects in the presence and absence of INH and/or L-GSH treatment (Figure 4I).

3.5. *Expression of PD-1 in the In Vitro Granulomas*

Programmed death 1 markers were measured with a conjugated horseradish peroxidase (HRP) anti-PD1 antibody with an anti-mouse FITC secondary antibody. Analysis of PD-1 expression was corrected with the average mean fluorescence of DAPI. Our lab observed a significant decrease in the expression of PD-1 in response to L-GSH treatment in both non-vaccinated and BCG-vaccinated subjects when compared to the sham-treated category (Figure 5A,B). When the granulomas were treated with PZA, there was a significant decrease in PD-1 expression in the non-vaccinated group among all categories when compared to the control group. Additionally, in response to L-GSH, there was a significant decrease in PD-1 expression to the 1/10 PZA + LGSH and 1/100 PZA + L-GSH when compared to their counterparts without L-GSH treatment (Figure 5C). In the BCG-vaccinated group, there was a decrease in PD-1 expression across all categories when compared to the control, and the addition of L-GSH further significantly reduced PD-1 expression when compared to the antibiotics without L-GSH (Figure 5D). Between the non-vaccinated and BCG-vaccinated sham-treated and L-GSH-treated groups, the BCG-vaccinated group had significantly less PD-1 expression in the sham-treated control (Figure 5E). When treated with PZA, the granulomas from BCG-vaccinated subjects demonstrated a significant decrease in PD-1 expression when compared to the non-vaccinated subjects, with the exception of 1/10 PZA + L-GSH and 1/100 PZA + L-GSH (Figure 5F). We observed an overall decreased expression of PD-1 from the granulomas of BCG-vaccinated subjects compared to the non-vaccinated group. The levels of PD-1 were measured in granulomas treated with INH in the presence and absence of L-GSH. In the non-vaccinated group, the addition of L-GSH leads to a significant decrease in PD-1 expression when compared to standalone antibiotics and the control. Additionally, there was a significant decrease of PD-1 expression in the 1/10 INH and 1/100 INH categories when compared to the control (Figure 5G). Similar to the non-vaccinated group, in the BCG-vaccinated group, L-GSH addition to the INH treatment groups leads to a significant decrease in PD-1 expression compared to the sham-treated and standalone antibiotic-treated categories. Additionally, there was a significant decrease of PD-1 in the MIC INH and 1/100 INH categories (Figure 5H). Importantly, the BCG-vaccinated group had significantly lower PD-1 levels at every category (INH treatment group in the presence or absence of L-GSH) when compared to the non-vaccinated group (Figure 5I).

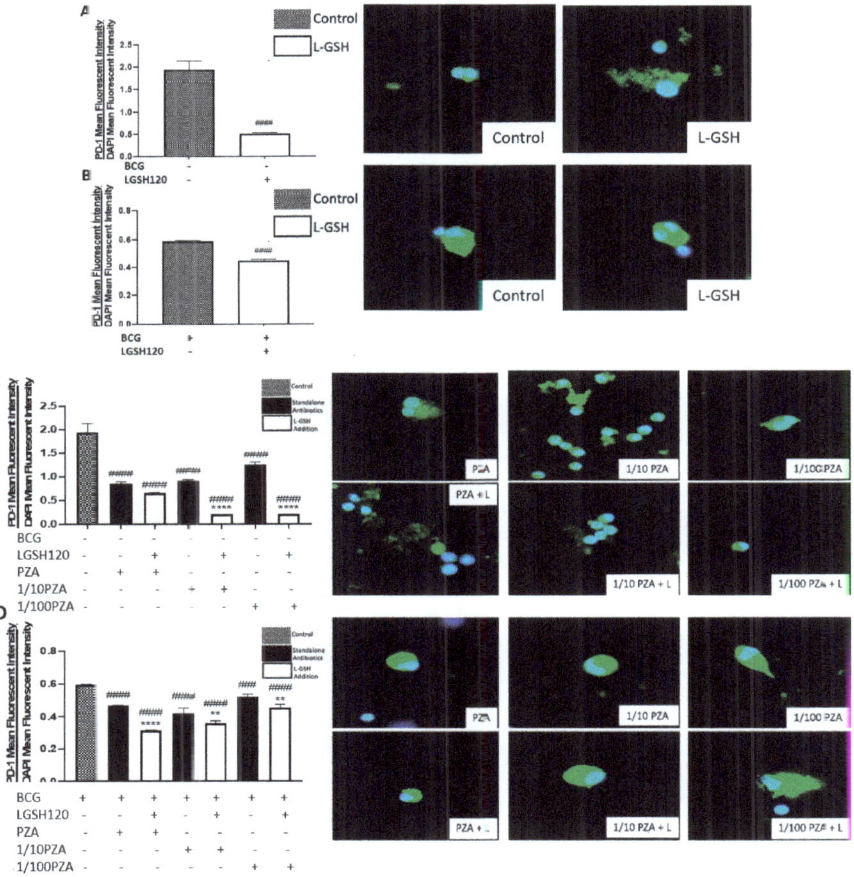

Figure 5. *Cont.*

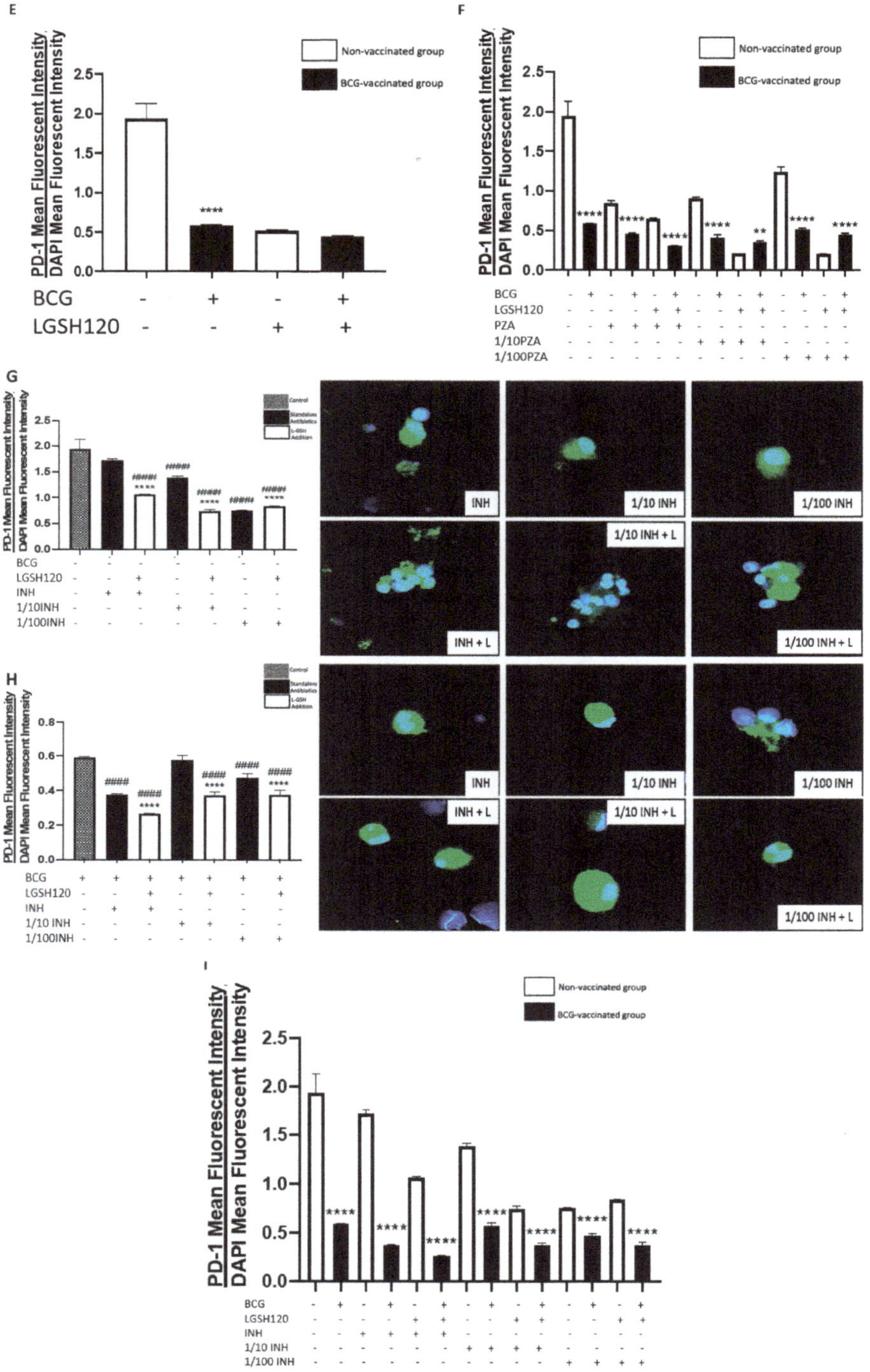

Figure 5. Expression of programmed death receptor 1 (PD-1) in PZA treated granulomas. PD-1 T cells

were measured with an anti-PD1 primary monoclonal antibody conjugated with HRP. A secondary anti-mouse PE-Cy5 antibody was added after blocking with 0.1% Triton X for 1 h. The nuclei of cells were stained with DAPI. The mean fluorescence of PD-1 was corrected with the mean DAPI fluorescence. (**A**) PD-1 mean fluorescence from non-vaccinated subjects with L-GSH treatment along with PD-1 fluorescent images; (**B**) PD-1 mean fluorescence from BCG vaccinated subjects with L-GSH treatment along with PD-1 fluorescent images; (**C**) PD-1 mean fluorescence with PZA, 1/10 PZA, and 1/100 PZA treatment in the presence or absence of L-GSH in non-vaccinated subjects; (**D**) PD-1 mean fluorescence with PZA, 1/10 PZA, and 1/100 PZA treatment in the presence or absence of L-GSH in BCG-vaccinated subjects; (**E**) PD-1 mean fluorescence in non-vaccinated and BCG-vaccinated groups with sham treatment and L-GSH treatment; (**F**) PD-1 mean fluorescence in non-vaccinated and BCG-vaccinated groups with PZA treatment. Data represents ± SE from experiments performed from 14 different subjects. An unpaired t-test with Welch corrections was used in Figure (**A**). Analysis of Figures B–F utilized a one-way ANOVA with Tukey test. The *p* value style consisted of 0.1234 as not significant, 0.0332 with one asterisk (*), 0.0021 with two asterisks (**), 0.0002 with three asterisks (***), and less than 0.0001 with four asterisks (****). A hash mark (#) indicates categories compared to the control, and an asterisk indicates categories compared to the previous category directly before it. (**G**) PD-1 mean fluorescence with INH, 1/10 INH, and 1/100 INH treatment in the presence or absence of L-GSH in non-vaccinated subjects; (**H**) PD-1 mean fluorescence with INH, 1/10 INH, and 1/100 INH treatment in the presence or absence of L-GSH in BCG-vaccinated subjects; (**I**) PD-1 mean fluorescence in non-vaccinated and BCG-vaccinated groups with INH treatment. Data represents ± SE from experiments performed from 14 different subjects. Analysis of figures utilized a one-way ANOVA with Tukey test. The *p* value style consisted of 0.1234 as not significant, 0.0332 with one asterisk (*), 0.0021 with two asterisks (**), 0.0002 with three asterisks (***), and less than 0.0001 with four asterisks (****). A hash mark (#) indicates categories compared to the control, and an asterisk indicates categories compared to the previous category directly before it.

3.6. Cytokine Responses in the In Vitro Granulomas

Using the sandwich Enzyme-linked Immune Sorbent Assay (ELISA) method, we measured the levels of TNF-α, a cytokine that causes the recruitment of other cells to form a granuloma, and IFN-γ, which enhances the effector functions of macrophages to control intracellular *M. tb* infection. These cytokines were measured from the supernatants of the terminated samples. TNF-α expression was significantly increased in the control and L-GSH treatment categories of the BCG-vaccinated group (Figure 6A). When the granulomas were treated with PZA, there was a general upwards trend in TNF-α production from the BCG vaccinated categories, especially in the lower concentrations (Figure 6B). However, due to complete bacterial clearance of the MIC treated categories, no further immunological responses were needed, which is why we ascertain that the TNF-α levels at the lower concentrations were low. There was a significant increase in IFN-γ in the BCG-vaccinated L-GSH treatment group (Figure 6C). PZA treatment in the presence and absence of L-GSH resulted in a significant increase of IFN-γ in the BCG-vaccinated group when compared to the non-vaccinated group (Figure 6D). Increased production of IFN-γ may be the mechanism by which BCG-vaccinated subjects display enhanced control over the bacteria. TNF-α and IFN-γ levels were measured in the supernatants of INH-treated granulomas. TNF-α production was significantly elevated in the control category and in the lower 1/10 INH + L-GSH and 1/100 INH + L-GSH120 categories in BCG vaccinated groups. Treatment with INH at MIC in the presence and absence of L-GSH resulted in complete inhibition in the growth of *M. tb*, leading to homeostasis in the immune response, therefore resulting in significantly lower TNF-α production in BCG-vaccinated subjects (Figure 6E). INH treatment of BCG-vaccinated granulomas in the presence or absence of L-GSH resulted in significantly higher levels of IFN-γ when compared to the non-vaccinated group (Figure 6F). This follows the same trend as those treated with PZA.

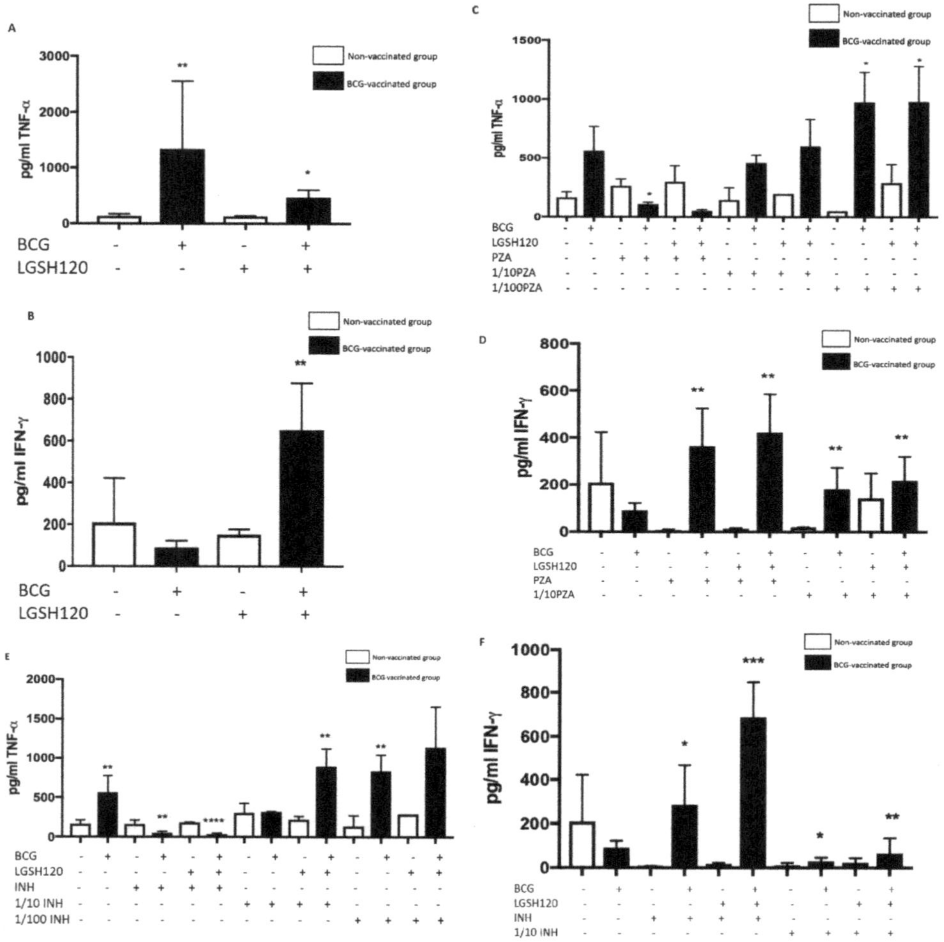

Figure 6. Cytokine responses in PZA and INH treated granulomas. (**A**) Tumor necrosis factor-α (TNF-*α*) levels from the supernatants of non-vaccinated and BCG-vaccinated sham-treated and L-GSH-treated granulomas at eight days post-infection; (**B**) TNF-*α* levels from the supernatants of non-vaccinated and BCG-vaccinated PZA treated granulomas at eight days post-infection; (**C**) interferon gamma (IFN-γ) levels from the supernatants of non-vaccinated and BCG-vaccinated sham-treated and L-GSH-treated granulomas at eight days post-infection; (**D**) IFN-γ levels from the supernatants of non-vaccinated and BCG-vaccinated PZA treated granulomas at eight days post-infection. Data represents ± SE from experiments performed from 14 different subjects. Analysis of figures utilized a one-way ANOVA with Tukey test. The *p* value style consisted of 0.1234 as not significant, 0.0332 with one asterisk (*), 0.0021 with two asterisks (**), 0.0002 with three asterisks (***), and less than 0.0001 with four asterisks (****). (E) TNF-α levels from the supernatants of non-vaccinated and BCG-vaccinated INH-treated granulomas at eight days post-infection; (**F**) IFN-γ levels from the supernatants of non-vaccinated and BCG-vaccinated INH-treated granulomas at eight days post-infection. Data represents ± SE from experiments performed from 14 different subjects. Analysis of figures utilized a one-way ANOVA with Tukey test. The *p* value style consisted of 0.1234 as not significant, 0.0332 with one asterisk (*), 0.0021 with two asterisks (**), 0.0002 with three asterisks (***), and less than 0.0001 with four asterisks (****).

3.7. Autophagy Levels in the In Vitro Granulomas

Most assays for autophagy modulators use the autophagy marker protein LC3B as the readout for autophagic activity. LC3B is a mammalian homolog of the yeast ATG8 protein, a ubiquitin-like protein that becomes lipidated and tightly associated with the autophagosomal membranes [19]. Autophagy levels were therefore measured by LC3B detection, a key marker for the autophagosome membrane structure. Increased levels of autophagy were observed in non-vaccinated individuals (Figure 7A). L-GSH treatment enhanced the levels of autophagy in BCG-vaccinated subjects (Figure 7B). In the granulomas of non-vaccinated subjects treated with PZA in the presence or absence of L-GSH, there was a significant increase of LC3B levels in all categories treated with L-GSH compared to the control and the categories without L-GSH treatment (Figure 7C). The same trend was found in the BCG-vaccinated subjects, with the exception of 1/10 PZA + L-GSH, with the only significant increase found when compared to the sham-treated group (Figure 7D). When comparing the non-vaccinated and BCG-vaccinated individuals directly, the BCG-vaccinated individuals showed a significant decrease in autophagy when compared to non-vaccinated individuals (Figure 7E). In response to PZA treatment with L-GSH, the LC3B levels were higher in the non-vaccinated group than the BCG-vaccinated group (Figure 7F). This may be a cell death mechanism to compensate for the low CD4 and high PD-1 in non-vaccinated subjects. In INH-treated granulomas, the presence of L-GSH lead to a significant increase in LC3B levels in the non-vaccinated group when compared to the sham-treated and the lone antibiotic-treated groups (Figure 7G). In BCG-vaccinated individuals, the MIC standalone INH showed a significant increase in autophagy compared to the control, and the addition of L-GSH to this category lead to a significant increase compared to the sham-treated and MIC standalone-treated groups (Figure 7H). When compared to the non-vaccinated group, there is less expression of LC3B in BCG-vaccinated individuals (Figure 7I). L-GSH treatment of granulomas from BCG-vaccinated subjects resulted in decreased expression of LC3B.

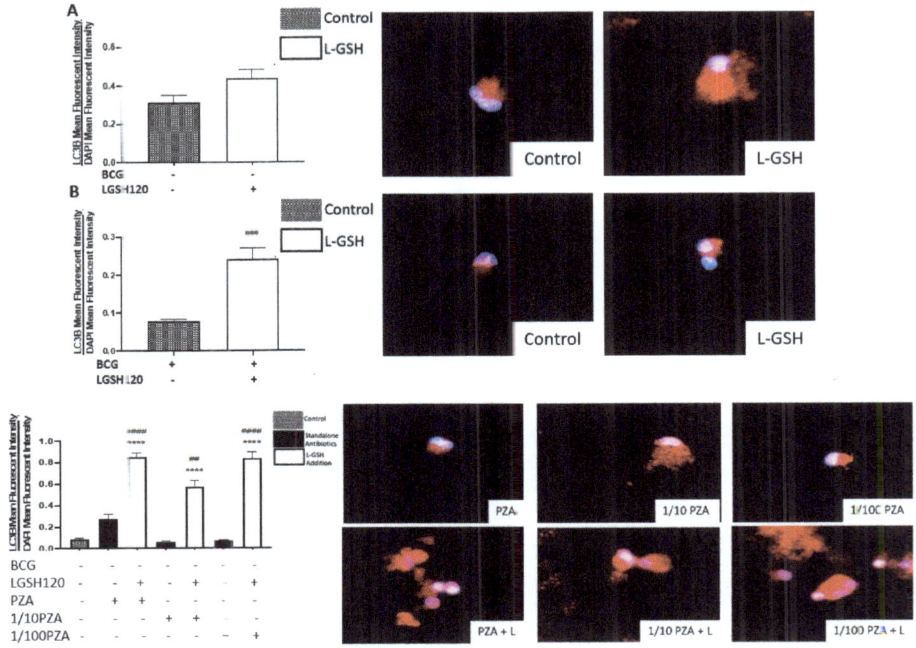

Figure 7. *Cont.*

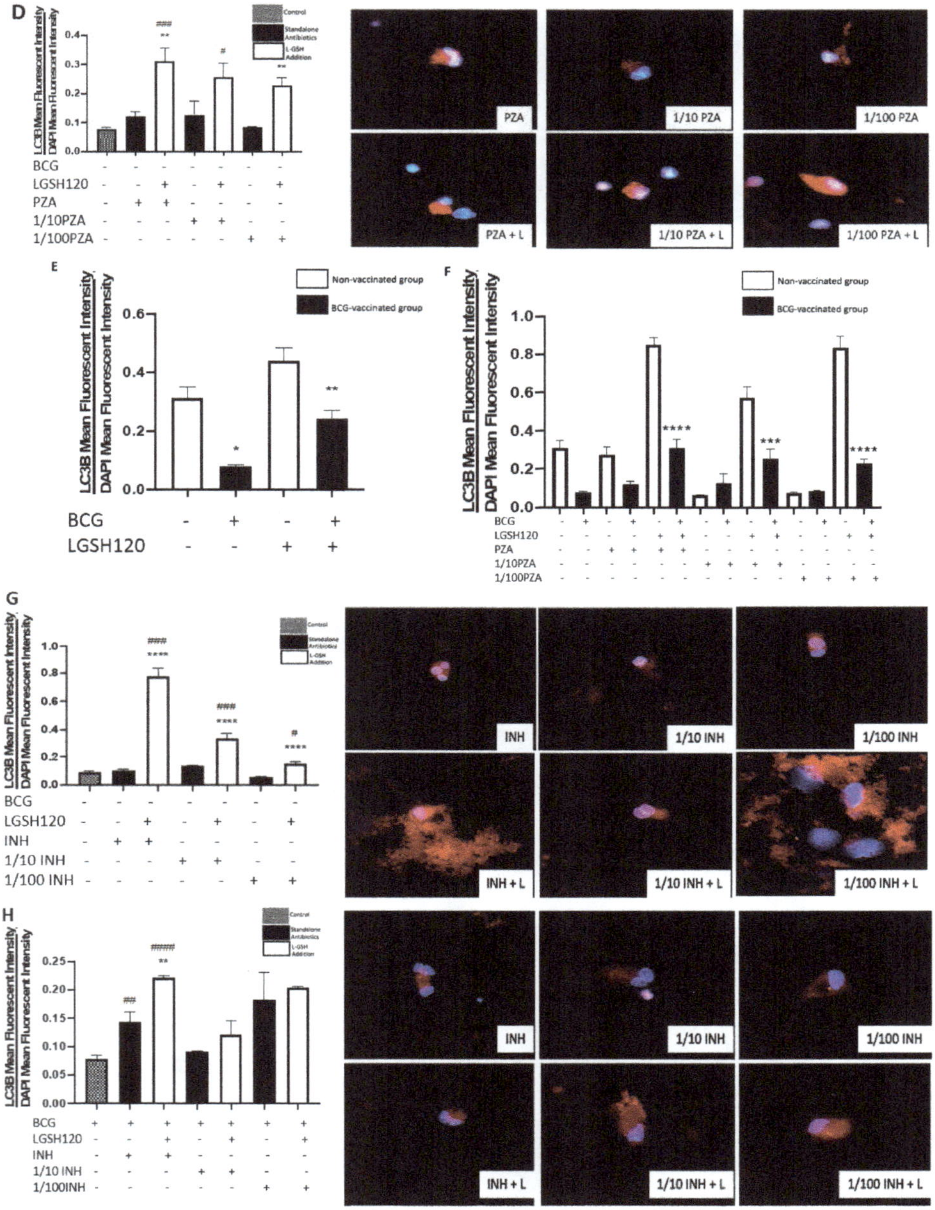

Figure 7. *Cont.*

I

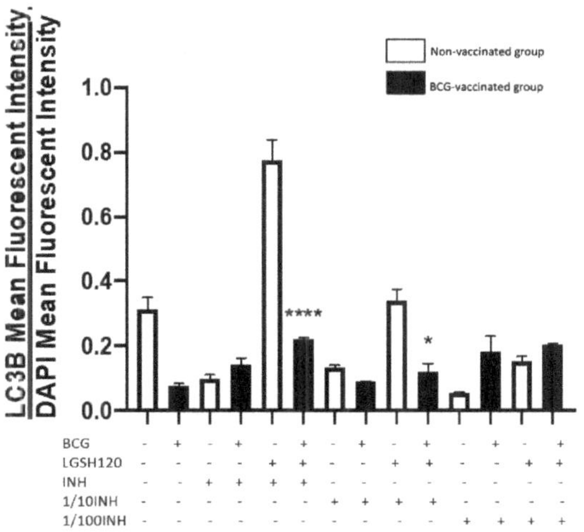

Figure 7. Autophagy levels in PZA- and INH-treated granulomas. Autophagy was measured with an anti-LC3B antibody, conjugated with PE-Cy5. The nuclei of cells were stained with DAPI. The mean fluorescence of LC3B was corrected with the mean DAPI fluorescence. (**A**) LC3B mean fluorescence from non-vaccinated subjects with L-GSH treatment, alongside LC3B fluorescent images; (**B**) LC3B mean fluorescence from BCG-vaccinated subjects with L-GSH treatment, alongside LC3B fluorescent images; (**C**) LC3B mean fluorescence with PZA, 1/10 PZA, and 1/100 PZA treatment in the presence or absence of L-GSH in non-vaccinated subjects; (**D**) LC3B mean fluorescence with PZA, 1/10 PZA and 1/100 PZA treatment in the presence or absence of L-GSH in BCG-vaccinated subjects; (**E**) LC3B mean fluorescence in non-vaccinated and BCG-vaccinated groups with sham treatment and L-GSH treatment; (**F**) LC3B mean fluorescence in non-vaccinated and BCG-vaccinated groups with PZA treatment. Data represents ± SE from experiments performed from 14 different subjects. An unpaired t-test with Welch corrections was used in Figure (**A**). Analysis of Figures B–F utilized a one-way ANOVA with Tukey test. The *p* value style consisted of 0.1234 as not significant, 0.0332 with one asterisk (*), 0.0021 with two asterisks (**), 0.0002 with three asterisks (***), and less than 0.0001 with four asterisks (****). A hash mark (#) indicates categories compared to the control, and an asterisk indicates categories compared to the previous category directly before it. (**G**) LC3B mean fluorescence with INH, 1/10 INH, and 1/100 INH treatment in the presence or absence of L-GSH in non-vaccinated subjects; (**H**) LC3B mean fluorescence with INH, 1/10 INH, and 1/100 INH treatment in the presence or absence of L-GSH in BCG-vaccinated subjects; (**I**) LC3B mean fluorescence in non-vaccinated and BCG-vaccinated groups with INH treatment. Data represents ± SE from experiments performed from 14 different subjects. Analysis of figures utilized a one-way ANOVA with Tukey test. The *p* value style consisted of 0.1234 as not significant, 0.0332 with one asterisk (*), 0.0021 with two asterisks (**), 0.0002 with three asterisks (***), and less than 0.0001 with four asterisks (****). A hash mark (#) indicates categories compared to the control, and an asterisk indicates categories compared to the previous category directly before it.

4. Discussion

Using *in vitro* granulomas derived from PBMCs, this study compares the differences in the immune responses against *M. tb* in both BCG-vaccinated and non-vaccinated subjects. The effects of GSH enhancement and the first line antibiotics (INH and PZA) in further enhancing the ability of immune cells to control *M. tb* infection in both the groups was also evaluated. This study also focuses on the mechanistic actions of PBMC-derived *in vitro* granuloma-like structures infected with the Erdman strain of *M. tb* from non-vaccinated subjects and BCG-vaccinated subjects to illustrate their immune response in a regulated environment. Our *in vitro* granulomas constituted cell types, such as macrophages, monocytes, and CD4 and CD8 T cells, all of which contribute to the immune responses necessary for granuloma formation.

We observed an active replication of *M. tb* to in the sham-treated granulomas, from the non-vaccinated group (Figure 1A). Furthermore, the addition of L-GSH led to a significant reduction in the viability of *M. tb* (Figure 1A). This confirms previous findings that indicate that L-GSH possesses antimycobacterial and immune-enhancing abilities. Furthermore, H&E staining was performed on these granuloma-like structures to visualize the morphology of these structures. The absence of L-GSH in the sham-treated category led to a formation of a sparse array of a granuloma, while L-GSH addition to this sample formed a larger, more robust granuloma-like structure (Figure 1A). This demonstrates the *M. tb* containment ability of L-GSH (Figure 1A). A similar trend was found in the BCG-vaccinated subjects. Notably, the mycobacterial count in the sham-treated control of the BCG-vaccinated group was 50% lower than the non-vaccinated subjects, and the addition of L-GSH lead to a further significant reduction in the bacterial load (Figure 1B). H&E staining demonstrated that the sham-treated and L-GSH-treated categories resulted in dense, robust granuloma-like structures, indicating the ability of BCG-vaccinated individuals to mount a robust immune response to hinder *M. tb* growth.

Findings from our study indicate that treatment of *in vitro* granulomas with PZA + L-GSH and INH+L-GSH resulted in a significant reduction in the bacterial load in the BCG-vaccinated samples compared to the non-vaccinated samples (Figure 1F,I). This evidence suggests that BCG vaccination enhances the ability of immune cells to effectively respond to GSH, PZA, and INH treatments, leading to containment and killing of *M. tb* (Figure 1E,F,I).

Our results also indicate that BCG-vaccinated individuals innately had higher levels of GSH when compared to the non-vaccinated group. Furthermore, L-GSH treatment resulted in a significant three log increase in the levels of GSH in BCG-vaccinated subjects when compared to non-vaccinated subjects (Figure 2A). PZA or INH treatments resulted in a significant increase in the levels of GSH in the granulomas from the BCG-vaccinated group when compared to the non-vaccinated group (Figure 2). Increased levels of GSH in the *in vitro* granulomas of BCG-vaccinated subjects further illustrate GSH's antimycobacterial activity, as indicated by low *M. tb* numbers in the CFU assays in the BCG-vaccinated group in comparison to the non-vaccinated group.

Macrophages, CD4 and CD8 T cells play an important role in the control of *M. tb* infection. Cytokines, such as TNF-α, interleukin-2 (IL-2), and IFN-γ, produced by macrophages and T cells can enhance the effector responses against *M. tb* infection. In order to understand the immune effector mechanisms, by which BCG-vaccinated versus non-vaccinated subjects contain *M. tb* infection, we first measured the expression levels of CD4 in the granulomas. We initially observed the mean fluorescence of CD4 in the granulomas cultured in the presence and/or absence of L-GSH, PZA and PZA. We observed a dramatic, significant increase in the expression of CD4 in the granulomas from the BCG-vaccinated group with L-GSH addition (Figure 3E), PZA addition at every concentration of the antibiotic in the presence and absence of L-GSH, and INH + L-GSH treatment at every concentration of the antibiotic (Figure 3F,I), indicating the efficacy of L-GSH, PZA, and INH to maintain the viability CD4 T cells (Figure 3). This evidence also demonstrates that there is less CD4 T cell exhaustion in BCG-vaccinated subjects.

$CD8^+$ T cells have also been shown to produce IFN-γ that can augment effector functions to control *M. tb* infection [20,21]. Importantly, $CD8^+$ T cells possess cytolytic and cytotoxic functions to kill *M. tb* infected cells via a granule-mediated function (using perforin, granzymes, and granulysin). CD8 expression was determined in the granulomas of non-vaccinated and BCG-vaccinated subjects. Higher CD8 expression was observed following L-GSH treatment in the granulomas from both non-vaccinated and BCG-vaccinated subjects (Figure 4E). The addition of the PZA and INH in the presence or absence of L-GSH resulted in an increase in the expression of CD8 in the BCG-vaccinated category when compared to the non-vaccinated group (Figure 4F,I). Based on this general upward trend in the expression of CD8 in BCG-vaccinated subjects, we conclude that individuals who have been vaccinated with BCG have not only higher CD4 T cells in the presence of L-GSH, but also elevated levels of CD8 T cells.

A programmed death receptor 1 (PD-1) is a type I transmembrane protein expressed in immune cells, such as T, B, and NK cells [22,23]. When it binds to its receptor, PD-L1, PD-1 strongly interferes with T cell receptor (TCR) signal transduction [22,23]. PD-1 is a negative regulator of activated T cells, and blockage of this path restores T cell functions [10,24]. For this reason, our lab quantified the levels of PD-1 in non-vaccinated and BCG-vaccinated subjects. We observed a significant decrease in the expression of PD1 in the sham control, PZA-treated, and INH-treated granulomas from BCG-vaccinated individuals when compared to the same categories in the non-vaccinated group (Figure 5F,I). The addition of L-GSH significantly reduced PD-1 expression in both non-vaccinated and BCG-vaccinated groups (Figure 5A,B). These results indicate that low PD-1 expression in BCG-vaccinated individuals leads to less T cell exhaustion during *M. tb* infection. Decreased PD-1 expression in BCG-vaccinated subjects, along with increased CD4 and CD8 expressions, resulted in improved killing and containment of *M. tb*.

Cytokine production is a key determinate in the containment of a mycobacterial infection. TNF-α released during initiation of host immune response against *M. tb* infection causes recruitment of immune cells to form a granuloma while IFN-γ enhances effector functions of macrophages to control an intracellular *M. tb* infection. While TNF-α is a vital cytokine in containing *M. tb*, overexpression of TNF-α has been connected to multiple autoimmune inflammatory diseases [24–26]. Hence modulation of this cytokine is necessary. BCG-vaccinated individuals had higher levels of TNF-α in the sham-treated category (Figure 6A). L-GSH has a modulatory effect on the production of TNF-α in the BCG-vaccinated subjects (Figure 6A). Treatment of granulomas from BCG-vaccinated subjects with lower concentrations of PZA and INH in the presence or absence of L-GSH led to higher TNF-α production (Figure 6C,E). In the MIC concentration for both PZA and INH, TNF-α levels were diminished in the BCG-vaccinated categories. This is because at MIC, the CFUs were undetectable and, hence, there is restoration of homeostasis in the immune responses. L-GSH, PZA, and INH treatments also significantly increased the production of IFN-γ in granulomas from BCG-vaccinated subjects when compared to the non-vaccinated subjects (Figure 6D,F). This increased production of IFN-γ may be the mechanism by which BCG-vaccinated subjects can control *M. tb* infection.

Recently, autophagy has also been recognized as an immune effector mechanism against intracellular pathogens [27,28]. We observed that BCG-vaccinated individuals have a significant decrease in the expression of LC3B when compared to non-vaccinated individuals in the presence and absence of L-GSH, PZA and INH (Figure 7F,I). Our results indicate that autophagy could be a compensatory mechanism by which non-vaccinated individuals combat an *M. tb* infection.

Our study findings illustrate that the addition of L-GSH to the antibiotic treatment elicited a significant improvement in the granulomatous responses against *M. tb* infection. Our results indicate that in non-vaccinated individuals there was increased survival of *M. tb*, low CD8 T cell counts, and higher expression of PD-1. The addition of L-GSH to granulomas from non-vaccinated individuals restored the viability of CD8 and CD4 T cells and promoted autophagy. In contrast to the non-vaccinated subjects, BCG-vaccinated subjects innately able to kill *M. tb* more effectively, and this was accompanied by increased levels of GSH, higher CD8 counts, and decreased expression of PD-1. The addition of L-GSH to granulomas from BCG-vaccinated subjects lead to increased CD4 T cell viability, increased TNF-α and IFN-γ production, decreased PD-1 expression, and diminished T cell exhaustion. Therefore, we believe that enhancing GSH by means of L-GSH supplementation, along with anti-TB treatment, would not only reduce toxicity and treatment duration, but can also enhance the host immune responses to combat an active infection to promote enhanced treatment compliance.

Author Contributions: V.V. conceived the study, developed study design, analyzed the data and prepared the manuscript. R.A. conducted the studies and drafted the manuscript. R.C., B.R., S.M., T.C., K.T., D.A., J.H., T.N. and G.T. provided technical assistance.

Acknowledgments: The authors appreciate the funding support from Your Energy Systems, Western University of Health Sciences and National Heart Blood Lung Institute at the National Institutes of Health (NIH) award 1R15HL143545-01A1 to conduct this study. We also appreciate the technical support of Wael Khamas, Albert Medina, and Edith Avitia. We thank Xiaoning Bi and Nissar Darmani for providing constructive feedback on this study. Finally, we sincerely thank the participants of this study for their time and involvement.

Conflicts of Interest: The authors declare no conflict of interest.

References

1. Tuberculosis (TB)—World Health Organization Fact Sheet. Available online: https://www.who.int/en/news-room/fact-sheets/detail/tuberculosis (accessed on 18 August 2019).
2. Ramakrishnan, L. Revisiting the role of the granuloma in tuberculosis. *Nat. Rev. Immunol.* **2012**, *12*, 352–366. [CrossRef] [PubMed]
3. Russell, D.G.; Cardona, P.J.; Kim, M.J.; Allain, S.; Altare, F. Foamy macrophages and the progression of the human tuberculosis granuloma. *Nat. Immunol.* **2009**, *10*, 943–948. [CrossRef] [PubMed]
4. Flynn, J.L.; Chan, J. Immunology of tuberculosis. *Annu. Rev. Immunol.* **2001**, *19*, 93–129. [CrossRef] [PubMed]
5. Guirado, E.; Schlesinger, L.S. Modeling the Mycobacterium tuberculosis Granuloma—The Critical Battlefield in Host Immunity and Disease. *Front. Immunol.* **2013**, *4*, 98. [CrossRef] [PubMed]
6. Forrellad, M.A.; Klepp, L.I.; Gioffre, A.; Sabio y Garcia, J.; Morbidoni, H.R.; de la Paz Santangelo, M.; Cataldi, A.A.; Bigi, F. Virulence factors of the Mycobacterium tuberculosis complex. *Virulence* **2013**, *4*, 3–66. [CrossRef] [PubMed]
7. Gautam, U.S.; Foreman, T.W.; Bucsan, A.N.; Veatch, A.V.; Alvarez, X.; Adekambi, T.; Golden, N.A.; Gentry, K.M.; Doyle-Meyers, L.A.; Russell-Lodrigue, K.E.; et al. In vivo inhibition of tryptophan catabolism reorganizes the tuberculoma and augments immune-mediated control of Mycobacterium tuberculosis. *Proc. Natl. Acad. Sci. USA* **2018**, *115*, E62–E71. [CrossRef] [PubMed]
8. Cooper, A.M.; Mayer-Barber, K.D.; Sher, A. Role of innate cytokines in mycobacterial infection. *Mucosal Immunol.* **2011**, *4*, 252–260. [CrossRef]
9. Van Crevel, R.; Ottenhoff, T.H.; van der Meer, J.W. Innate immunity to Mycobacterium tuberculosis. *Clin. Microbiol. Rev.* **2002**, *15*, 294–309. [CrossRef]
10. Okiyama, N.; Katz, S.I. Programmed cell death 1 (PD-1) regulates the effector function of CD8 T cells via PD-L1 expressed on target keratinocytes. *J. Autoimmun.* **2014**, *53*, 1–9. [CrossRef]
11. Treatment for TB Disease Treatment TB CDC. Available online: https://www.cdc.gov/tb/topic/treatment/tbdisease.htm (accessed on 18 August 2019).
12. Sacksteder, K.A.; Protopopova, M.; Barry, C.E., 3rd; Andries, K.; Nacy, C.A. Discovery and development of SQ109: A new antitubercular drug with a novel mechanism of action. *Future Microbiol.* **2012**, *7*, 823–837. [CrossRef]
13. Andersen, P.; Doherty, T.M. The success and failure of BCG—Implications for a novel tuberculosis vaccine. *Nat. Rev. Microbiol.* **2005**, *3*, 656–662. [CrossRef] [PubMed]
14. Lagman, M.; Ly, J.; Saing, T.; Kaur Singh, M.; Vera Tudela, E.; Morris, D.; Chi, P.T.; Ochoa, C.; Sathananthan, A.; Venketaraman, V. Investigating the causes for decreased levels of glutathione in individuals with type II diabetes. *PLoS ONE* **2015**. [CrossRef] [PubMed]
15. Teskey, G.; Cao, R.; Islamoglu, H.; Medina, A.; Prasad, C.; Prasad, R.; Sathananthan, A.; Fraix, M.; Subbian, S.; Zhong, L.; et al. The Synergistic Effects of the Glutathione Precursor, NAC and First-Line Antibiotics in the Granulomatous Response Against Mycobacterium tuberculosis. *Front. Immunol.* **2018**. [CrossRef] [PubMed]
16. Cao, R.; Islamoglu, H.; Teskey, G.; Gyurjian, K.; Abrahem, R.; Onajole, O.; Lun, S.; Bishai, W.; Kozikowski, A.; Fraix, M.P.; et al. The preclinical candidate indole-2-carboxamide improves immune responses to Mycobacterium tuberculosis infection in healthy subjects and individuals with type 2 diabetes. *Int. Microbiol.* **2019**. [CrossRef] [PubMed]
17. Islamoglu, H.; Cao, R.; Teskey, G.; Gyurijian, K.; Lucar, S.; Fraix, M.; Sathananthan, A.; Venketaraman, V. Effects of Readisorb L-GSH in altering granulomatous responses against Mycobacterium tuberculosis infection. *J. Clin. Med.* **2018**, *7*, 40. [CrossRef] [PubMed]
18. Cao, R.; Teskey, G.; Islamoglu, H.; Gutierrez, M.; Salaiz, O.; Munjal, S.; Fraix, M.P.; Sathananthan, A.; Nieman, D.C.; Venketaraman, V. Flavonoid Mixture Inhibits Mycobacterium tuberculosis Survival and Infectivity. *Molecules* **2019**, *24*, 851. [CrossRef] [PubMed]

19. Tanida, I.; Ueno, T.; Kominami, E. LC3 and Autophagy. *Methods Mol. Biol.* **2008**, *445*, 77–88. [CrossRef]
20. Jasenosky, L.D.; Scriba, T.J.; Hanekom, W.A.; Goldfeld, A.E. T cells and adaptive immunity to Mycobacterium tuberculosis in humans. *Immunol. Rev.* **2015**, *264*, 74–87. [CrossRef]
21. Lin, P.L.; Flynn, J.L. CD8 T cells and Mycobacterium tuberculosis infection. *Semin. Immunopathol.* **2015**, *37*, 239–249. [CrossRef]
22. Arasanz, H. Gato-Canas, M.; Zuazo, M.; Ibanez-Vea, M.; Breckpot, K.; Kochan, G.; Escors, D. PD1 signal transduction pathways in T cells. *Oncotarget* **2017**, *8*, 51936–51945. [CrossRef]
23. Amarnath, S.; Mangus, C.W.; Wang, J.C.; Wei, F.; He, A.; Kapoor, V.; Foley, J.E.; Massey, P.R.; Felizardo, T.C.; Riley, J.L.; et al. The PDL1-PD1 axis converts human TH1 cells into regulatory T cells. *Sci. Transl. Med* **2011**. [CrossRef] [PubMed]
24. Tian, T.; Li, X.; Zhang, J. mTOR Signaling in Cancer and mTOR Inhibitors in Solid Tumor Targeting Therapy. *Int. J. Mol. Sci.* **2019**, *20*, 755. [CrossRef] [PubMed]
25. Desplat-Jego, S.; Burkly, L.; Putterman, C. Targeting TNF and its family members in autoimmune/inflammatory disease. *Mediat. Inflamm.* **2014**. [CrossRef] [PubMed]
26. Esposito, E.; Cuzzocrea, S. TNF-alpha as a therapeutic target in inflammatory diseases, ischemia-reperfusion injury and trauma. *Curr. Med. Chem.* **2009**, *16*, 3152–3167. [CrossRef] [PubMed]
27. Harris, J.; De Haro, S.A.; Master, S.S.; Keane, J.; Roberts, E.A.; Delgado, M.; Deretic, V. T helper 2 cytokines inhibit autophagic control of intracellular Mycobacterium tuberculosis. *Immunity* **2007**, *27*, 505–517. [CrossRef] [PubMed]
28. McEwan, D.G. Host-pathogen interactions and subversion of autophagy. *Essays Biochem.* **2017**, *61*, 687–697. [CrossRef] [PubMed]

Journal of
Clinical Medicine

MDPI

Article

Effect of Iron Supplementation on the Outcome of Non-Progressive Pulmonary *Mycobacterium tuberculosis* Infection

Afsal Kolloli †, Pooja Singh †, G. Marcela Rodriguez and Selvakumar Subbian *

The Public Health Research Institute Center of New Jersey Medical School, Rutgers University, Newark, NJ 07103, USA

* Correspondence: subbiase@njms.rutgers.edu; Tel.: +1-973-854-3226

† These authors contributed equally.

Received: 1 July 2019; Accepted: 31 July 2019; Published: 2 August 2019

check for updates

Abstract: The human response to *Mycobacterium tuberculosis* (Mtb) infection is affected by the availability of iron (Fe), which is necessary for proper immune cell function and is essential for the growth and virulence of bacteria. Increase in host Fe levels promotes Mtb growth and tuberculosis (TB) pathogenesis, while Fe-supplementation to latently infected, asymptomatic individuals is a significant risk factor for disease reactivation. However, the effect of Fe-supplementation on the host immunity during latent Mtb infection remains unclear, due partly to the paucity in availability of animal models that recapitulate key pathophysiological features seen in humans. We have demonstrated that rabbits can develop non-progressive latency similar to infected humans. In this study, using this model we have evaluated the effect of Fe-supplementation on the bacterial growth, disease pathology, and immune response. Systemic and lung Fe parameters, gene expression profile, lung bacterial burden, and disease pathology were determined in the Mtb-infected/Fe- or placebo-supplemented rabbits. Results show that Fe-supplementation to Mtb-infected rabbits did not significantly change the hematocrit and Hb levels, although it elevated total Fe in the lungs. Expression of selected host iron- and immune-response genes in the blood and lungs was perturbed in Mtb-infected/Fe-supplemented rabbits. Iron-supplementation during acute or chronic stages of Mtb infection did not significantly affect the bacterial burden or disease pathology in the lungs. Data presented in this study is of significant relevance for current public health policies on Fe-supplementation therapy given to anemic patients with latent Mtb infection.

Keywords: tuberculosis; latent infection; pulmonary; rabbit; *Mycobacterium tuberculosis*; iron supplementation; pathology; immune response; gene expression; Perls' stain

---

1. Introduction

About a third of the world's population is estimated to be infected with *Mycobacterium tuberculosis* (Mtb), the etiological agent of tuberculosis (TB) [1,2]. In more than 90% of Mtb-infected individuals, the immune response controls the infection, with Mtb persisting in a quiescent state [1]. These latentlyinfected (LTBI) individuals can active Mtb replication and symptomatic disease upon host immune compromising conditions [1]. Despite the role of host immunity in controlling Mtb infection, the specific components that contribute to the establishment and/or reactivation of LTBI remain poorly understood [3,4]

The human response to Mtb infection is affected by the availability of iron (Fe), which is necessary for the proper function of the immune system [5]. About a quarter of the total Fe in the body is present/stored in the hepatocytes and macrophages, while more than 60% is bound by hemoglobin, which is essential for the transfer and transport of oxygen [5]. About 5% of Fe in the body exist in proteins involved in respiration and energy metabolism. Thus, Fe homeostasis is important for normal cell growth and replication. In addition, several Fe-containing enzymes and proteins such as myeloperoxidase, NADPH oxidase, nitric oxide synthase, indole-amine 2,3 dioxygenase and lipoxygenase are directly involved in the host immune response to infection [5].

The World Health Organization (WHO) has identified Fe deficiency as the most common nutritional disorder worldwide and has developed a strategy to eradicate anemia in TB-endemic countries that includes increasing Fe intake through food fortification programs (https://www.who.int/nutrition/publications/en/ida_assessment_prevention_control.pdf). However, this program does not screen individuals for prior exposure to Mtb before starting therapy. This is a matter of concern because Fe is also essential for the growth and virulence of Mtb. Since host Fe homeostasis is tightly regulated, Mtb has to compete with the host cell to acquire Fe. As an intracellular pathogen, Mtb thrives mostly in macrophages, which contain high levels of cytoplasmic Fe. First, the host produces Fe-binding proteins, such as transferrin, ferritin and lactoferrin, in order to deprive invading pathogens of Fe and macrophages sequester these host proteins [6,7]. Second, macrophages degrade senescent erythrocytes, which release Fe-containing heme [8]. However, cytoplasmic Fe is not readily accessible to the intracellular Mtb that survive in the phagosomes. To acquire host Fe, Mtb produces Fe-chelating siderophores, which can scavenge Fe from non-heme host proteins and also facilitate the accumulation of Fe within the phagosomes of infected macrophages [8]. In infected host cells, Mtb-containing phagosomes can fuse with early endosomes, which contain transferrin, a protein that binds and transports Fe in the blood, allowing Mtb to acquire transferrin-bound Fe [9]. Increases in host Fe levels promote Mtb growth and TB pathogenesis [10–12]. This is evidenced by a recent study, which showed that in Fe-deficient microenvironments, such as within hypoxic granulomas, Mtb adapts to a non-replicating persistent state [13]. In addition, Esx-3, one of the type VII secretion systems of Mtb that is involved in Fe-acquisition has also been suggested to interfere with the host functions that restrict Fe-availability [14]. Moreover, Mtb mutant for *esx3* was impaired for Fe-uptake [14]. Similarly, Mtb mutant for *mbtK*, a mycobactin synthase gene involved in siderophore production, is defective for Fe-acquisition and showed attenuated growth in vitro and in infected mouse lungs [15]. These findings clearly show that Mtb utilizes several mechanisms to acquire Fe from the host and highlight the importance of Fe acquisition for the growth and virulence of Mtb. Indeed, treatment of Mtb-infected mice with Fe promoted bacterial growth in tissues and Fe-supplementation to anemic individuals has been associated with reactivation of LTBI [16–18]. Thus, providing Fe-supplementation therapy to asymptomatic LTBIs can be a significant risk factor for Mtb reactivation to symptomatic disease. In order to develop new and/or better public health strategies to control TB in high-burden countries, it is necessary to understand the role of host Fe homeostasis in host-pathogen interaction(s) during Mtb infection.

Although Fe is known to promote Mtb growth, the effect of Fe-supplementation on the course of initial Mtb infections and its influence on the host immunity to infection is unclear [10]. Commonly used animal models, such as mice fail to recapitulate key pathophysiological features associated with LTBI. Previously, we have demonstrated that Mtb CDC1551-infected rabbits can develop LTBI spontaneously, as seen in humans [19]. In this study, we have used this LTBI rabbit model to investigate the effect of moderate Fe-supplementation on the immune response and outcome of pulmonary Mtb infection.

2. Materials and Methods

2.1. Bacteria and Chemicals

Mycobacterium tuberculosis CDC1551 (Mtb CDC1551) strain was obtained from Dr. Thomas Shinnick (Centers for Disease Control (CDC), Atlanta, GA), grown in Middlebrook 7H9 medium supplemented with 10% OADC enrichment (Difco BD, Franklin Lakes, NJ, USA), aliquoted and stored frozen at −80 °C. To prepare the inoculum for rabbit aerosol infection, stock vials were thawed and used as described previously [19]. All chemicals were purchased from Sigma (Sigma-Aldrich, St. Louis, MO, USA), unless mentioned otherwise.

2.2. Experimental Design

Based on our previous experience with the rabbit model of LTBI and on the literature for Fe-supplementation therapy, we set out to investigate: (a) The effect of Fe-supplementation during "acute phase" of Mtb infection (i.e., from day 1 until 8 weeks post-infection) and (b). the effect of Fe-supplementation of rabbits with a pre-established infection, i.e., starting at 8 until 16 weeks post-infection, when untreated rabbits usually develop a non-progressive "chronic phase of infection" that ultimately leads to LTBI [19].

2.3. Rabbit Infection and Treatment

Fifty-eight ($n = 58$) specific pathogen free, female New Zealand white rabbits (*Oryctolagus cuniculus*) of ~2.5 kg body weight (Covance Research Products, Denver, PA, USA) were exposed to Mtb CDC1551-containing aerosols using a "nose-only" delivery system to deliver ~3 $\log_{10}$ CFU into the lungs, as described previously [19]. At 3 hours post-exposure (T = 0), six rabbits were euthanized, lungs were harvested, and serial dilutions of the lung homogenates were plated on Middlebrook 7H11 agar media (Difco BD, Franklin Lakes, NJ, USA) to determine the number of bacterial colony-forming units (CFU). Starting at day 1 or at 8 weeks post-infection, groups of rabbits ($n = 6$ per group per time point) were randomly allocated and treated with 25 mg Fe-dextran (in 0.5 mL water) or placebo (0.5 mL sterile dextran in water) injected intra-muscularly (biceps femoris), as described previously [20]. Fe-supplementation was given 3 days a week for 8 weeks (i.e., either from day 1 until 8 weeks post-infection for acute infection, or from 8 to 16 weeks post-infection for chronic infection) (Supplementary Figure S1). This dose of Fe-dextran III (75 mg/week) is equivalent to the recommended pediatric human dose, which is non-toxic and well tolerated in rabbits [20]. The animals were given food and water ad-libitum and, weighed periodically. At T = 0 (3 h), 4, 8, 12, and 16 weeks post-infection, groups of rabbits ($n = 6$) were euthanized and organs were harvested to enumerate Mtb CFU, histology and for total host RNA isolation. About 40% (by weight) of the lung was used for preparing homogenates for Mtb CFU assays. Four animals were used as uninfected controls for Fe estimation and gene expression analysis in the blood and lung tissues. Lung tissues for RNA isolation were snap-frozen at −80 °C immediately after removal. Tissue sections for histology were fixed in neutral formalin. Blood was collected in heparinized tubes and plasma was separated by centrifugation and used in iron estimation assays. The blood pellet was used for total RNA isolation. All animal procedures were performed as per the approved procedures of the Rutgers University Institutional Animal Care and Use Committee.

2.4. Histology Staining

Formalin-fixed lung portions were paraffin embedded and cut into 5 μm sections for staining with Hematoxylin–eosin (H&E) to visualize leukocytes, or Perls' staining method for iron deposition as reported previously [21]. The stained sections were analyzed using a Nikon Microphot DXM 1200C microscope and photographed using NIS-Elements software (Nikon Instruments Inc., Melville, NY, USA). Visualization of Mtb in infected rabbit lung sections was performed by immunohistochemistry (IHC) using anti-Mtb antibody, as reported previously [22]. The IHC-stained sections were analyzed and photographed using an EVOS FL fluorescence microscope (Thermo Fisher Scientific, Pittsburg, PA, USA).

2.5. Measurement of Blood Parameters

Hematocrit was performed manually using heparinized capillary tubes (Fisher Scientific, Pittsburg, PA, USA), as described previously [23]. Hemoglobin (Hb) was measured by using Hemoglobin Assay Kit following the manufacturer's instructions (Abnova, Walnut, CA, USA).

2.6. Measurement of Plasma and Lung Tissue Iron

Total iron-binding capacity (TIBC), total iron, and percentage transferrin saturation (%Tf) in plasma and lung homogenates of uninfected or Mtb CDC1551-infected rabbits with or without Fe-treatment was determined using a colorimetric assay (TIBC and Serum Iron Assay Kit) following the manufacturer's instructions (BioVision, Milpitas, CA, USA). This experiment was performed in duplicate with samples from three animals per time point per condition.

2.7. RNA Isolation from Rabbit Lung and Blood

Total host RNA was isolated as described earlier [24]. Briefly, lung tissue and whole blood cell pellet were homogenized with 10× volume (wt/vol) of TRI reagent (Molecular Research Center, Cincinatti, OH, USA). The homogenates were extracted with 0.3 volumes (vol/vol) of BCP solution and the aqueous phase was mixed with an equal volume of 70% ethanol and passed through mini spin columns (Qiagen Inc, Germantown, MD, USA). Following on-column digestion with DNaseI, and subsequent washings, RNA was eluted with sterile water. The quality and quantity of the purified RNA was assessed in a NanoDrop instrument (NanoDrop products, Wilmington, DE, USA).

2.8. Quantitative Real-Time PCR Analysis (qPCR)

Total RNA was reverse transcribed into cDNA using the AffinityScript QPCR cDNA Synthesis Kit following instructions of the manufacturer (Agilent Technologies Inc., Santa Clara, CA). Quantitative PCR (qPCR) experiments were performed on a Stratagene Mx3005p machine (Agilent Technologies, Inc. Santa Clara, CA, USA) with cDNA and site-specific oligonucleotide primers of target genes, using Brilliant III Ultra-Fast SYBR®Green QPCR Master Mix, according to the product instructions (Agilent Technologies Inc., Santa Clara, CA, USA). No SYBR and no cDNA control samples were included in one of the triplicate assays for each experimental time point. Housekeeping gene *GAPDH* was included to normalize the levels of expression of test genes. Threshold cycle value (C_t) was determined using MxPro4000 software and fold-change in gene expression was calculated from the formula $2^{-\Delta\Delta C_t}$, where ΔC_t is the difference in C_t between the test gene and *GAPDH* and $\Delta\Delta C_t$ is the difference between test and control conditions. Each experiment was repeated at least three times with cDNA from 3 to 6 animals per experimental time point per group. Supplementary Table S1 lists the description of tested genes and primer sequences used in qPCR experiments.

2.9. Statistical Analysis

Statistical analysis was performed on data by one-way ANOVA followed by Tukey's multiple test comparison using GraphPad Prism-5 (GraphPad Software, La Jolla, CA, USA) and the mean ± standard deviation values were plotted as graphs. For all the experimental data, $p \leq 0.05$ was considered statistically significant.

3. Results

3.1. Iron Parameters in Fe-Supplemented Rabbits during Acute or Chronic Mtb Infection

The effect of Fe-supplementation on the hematocrit and hemoglobin (Hb) content of rabbits during Mtb-infection was measured at 4 and 8 weeks (acute) or 12 and 16 weeks (chronic) post-infection/treatment.

An average hematocrit level of 42% and Hb content of ~14 g/L was observed at all time-points (Supplementary Figure S2), and no significant difference in either the hematocrit or Hb levels was observed in Mtb-infected rabbits with Fe- or placebo supplementation.

3.2. Systemic and Lung Fe Levels in Mtb-Infected Rabbits with or without Fe-Supplementation

The total Fe binding capacity (TIBC), which is an indirect measure of free Fe, total Fe, and the percent of Fe-saturated transferrin (%Tf) were measured in rabbit plasma and lung homogenate.

In the plasma, Mtb-infected rabbits with Fe-supplementation had significantly reduced TIBC, accompanied by elevated total Fe and %Tf saturation, compared to placebo-treated animals at both acute and chronic stages of infection (Figure 1A–C). In the lungs, no significant differences in TIBC and %Tf saturation levels were observed between animals, irrespective of their Fe-supplementation status during the acute phase of infection (Figure 2A,C). However, a higher Fe level was observed in the lungs of Mtb-infected/Fe-supplemented animals, compared to placebo-treated animals during acute (8 weeks) and chronic (16 weeks) post infection (Figure 2B). Fe-supplementation during chronic Mtb infection significantly increased the lung TIBC only at 16 weeks, compared to the placebo-treated rabbits (Figure 2A). Similarly, the Fe-supplemented rabbits also had a significantly higher Fe level in the lungs at 4 and 12 weeks post infection, compared to the placebo-supplemented animals (Supplementary Figure S3).

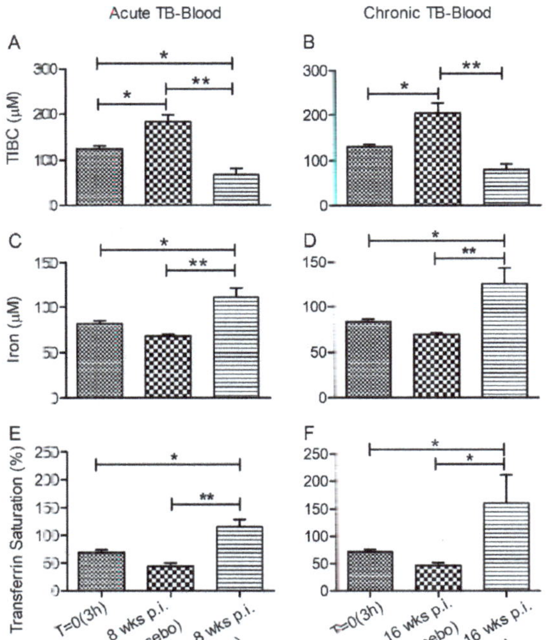

Figure 1. Effect of Fe-supplementation on plasma iron parameters of rabbits at 8 weeks (**A**,**C**,**E**) or 16 weeks (**B**,**D**,**F**) post-Mtb infection. Total iron binding capacity (TIBC) (top), total iron (middle), and percent transferrin saturation (bottom) were determined in plasma samples of Mtb-infected and placebo- or Fe-treated rabbits at the conclusion of treatment (i.e., 8 weeks or 16 weeks post-infection) Data was analyzed by one-way Anova with Tukey's multiple comparison test. Values plotted are mean +/– sd with $n = 4$ per group per time point. * $p < 0.05$; ** $p < 0.01$.

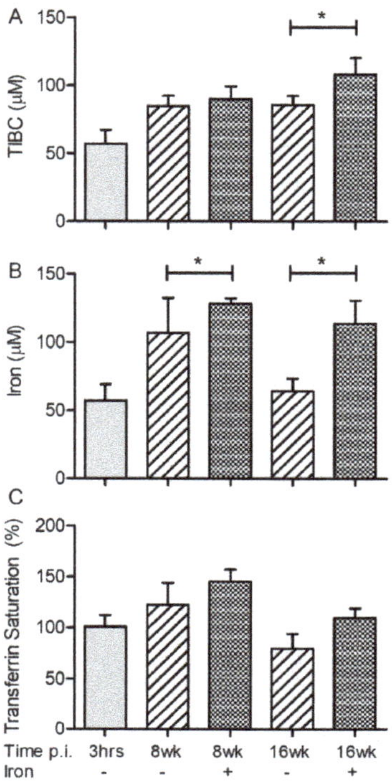

Figure 2. Effect of Fe-supplementation on the lung iron parameters of rabbits at 8 and 16 weeks post infection. TIBC (top; **A**), total iron (middle; **B**) and percent transferrin saturation (bottom; **C**) were determined in the homogenates of Mtb-infected and placebo (-) or Fe (+) treated rabbits. Data was analyzed by one-way Anova with Tukey's multiple comparison test. Values plotted are mean +/– sd with $n = 4$ per group per time point. * $p < 0.05$.

3.3. Fe-Supplementation Does Not Affect Systemic Host Iron-Responsive Gene Expression during Mtb Infection

The effect of Fe-supplementation on the systemic expression of host iron-responsive genes, including *HFE1*, *HFE2* (*HJV*), *HFE3* (*TFR2*), *HAMP*, *SLC40A1* (*FPN1*), *SLC11A2* (*NRAMP2*), *HMOX1*, *FTH1*, *LCN2*, *BMP6*, and *NFE2L2* (*NRF2*) was determined in Mtb-infected rabbit blood and lungs.

These genes are selected for their role in systemic Fe homeostasis as well as host response to Fe in the system (Supplementary Table S2). Among these genes, *HFE1* and *HFE2* interact with the transferrin receptor (TFR) and transferrin to modulate Fe transportation and availability. Transferrin also controls expression of hepcidin (HAMP), the central regulator of systemic Fe [25,26]. Moreover, HFE2 is a co-receptor for BMP, which drives HAMP expression through the SMAD signaling pathway [27]. The Fe exporter, FPN, mediates export of Fe recycled in macrophages from erythrocytes and Fe absorbed from the diet by enterocytes. HAMP binds to FPN and induces its internationalization and subsequent degradation, thereby reducing Fe release from macrophages and enterocytes [28]. FTH1 is a subunit of ferritin, an Fe-storage protein [29], and HMOX is an enzyme involved in heme catabolism, both of which are regulated by host Fe status [30]. NRF2, a regulator of the antioxidant response, also regulates FTH1 and FPN to maintain cellular Fe homeostasis [31]. LCN2 is an immune protein that sequesters Fe-loaded siderophores and blocks bacterial Fe acquisition from infected host cells [32].

Consistent with activation of nutritional immunity, expression of all the tested Fe-responsive genes was significantly up-regulated in the blood of Mtb-infected rabbits at 8 weeks (acute) (Figure 3A) and 16 weeks (chronic) (Figure 3B) post infection, compared to the uninfected animals. Similar trend in the expression of these genes was noted in rabbit blood at 4 (acute) and 12 (chronic) weeks post infection (Supplementary Figure S4). However, no significant differential expression was observed in the blood between placebo- and Fe- supplemented Mtb-infected rabbits at any of the tested time points (Figure 4A–D). These results suggest that Mtb infection induces the expression of host Fe-responsive genes systemically and that Fe-supplementation does not lead to further up-regulation of these genes.

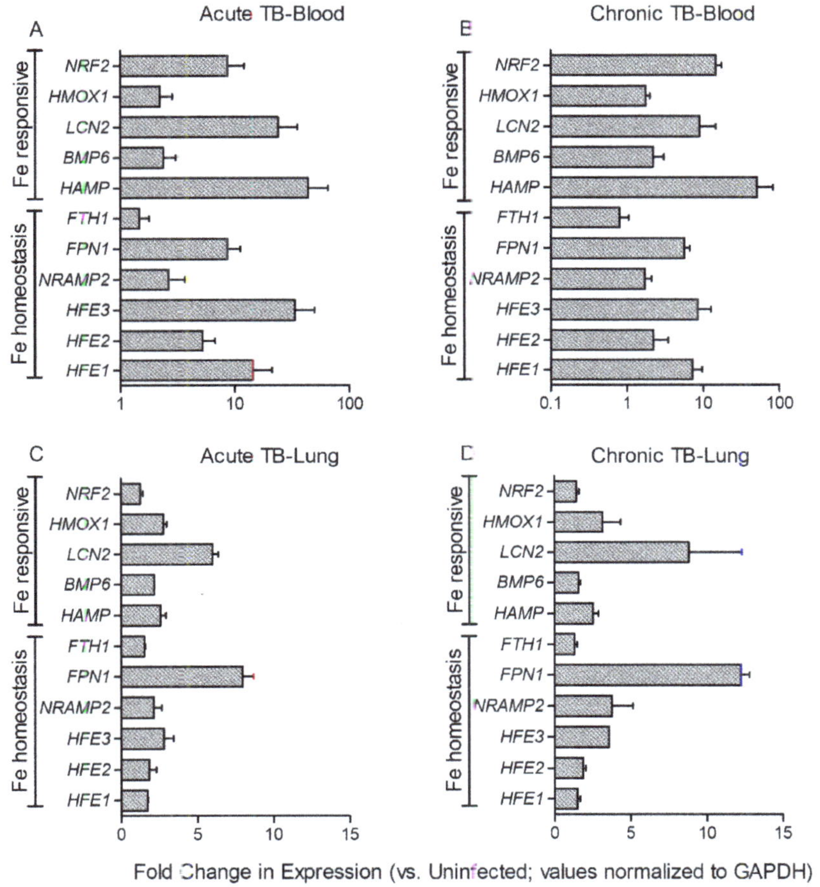

Figure 3. Expression of iron-responsive genes in *Mycobacterium tuberculosis* (Mtb)-infected rabbits at 4- and 12-weeks post infection. Data shown are expression of target genes in the blood (**A**,**B**) or lung (**C**,**D**) during acute (8 weeks; **A**,**C**) or chronic (16 weeks; **B**,**D**) stages of infection. The gene expression levels in Mtb-infected animals was calibrated with the corresponding levels in uninfected rabbits. Host house-keeping gene (*GAPDH*) expression was used to normalize the level of target gene expression. Data was analyzed by one-way Anova with Tukey's multiple comparison test. Values plotted are mean +/− sd with $n = 4$ per group per time point. All tested genes were statistically significant in Figure 3A–D

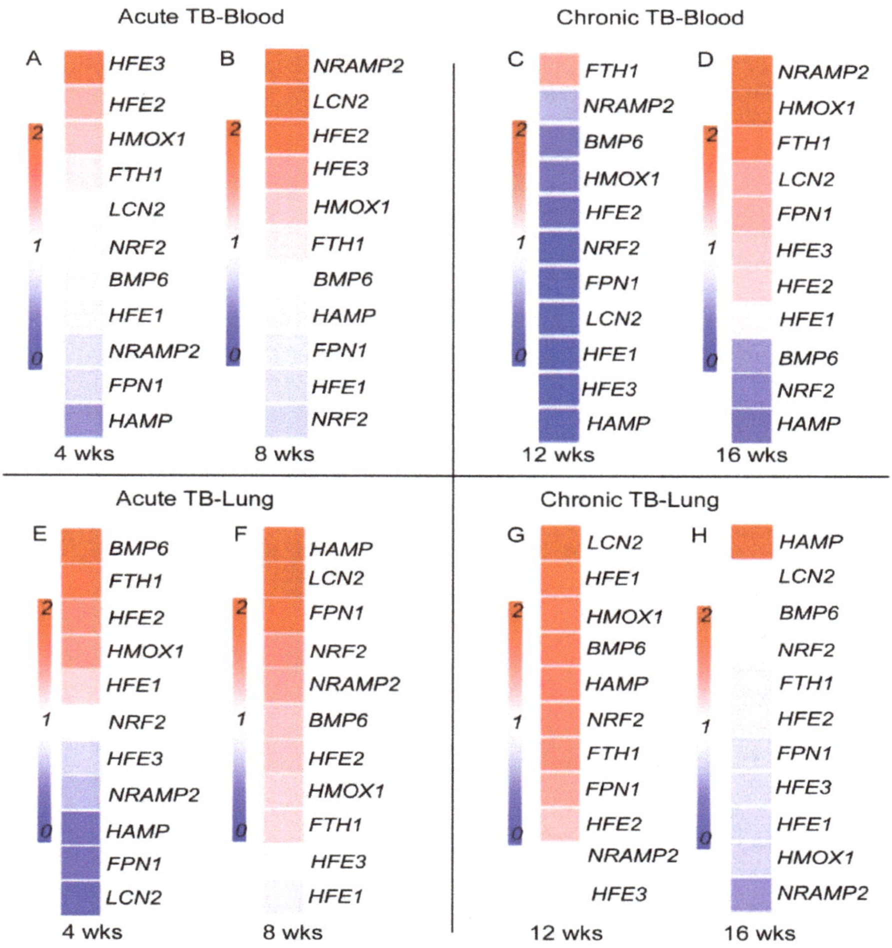

Figure 4. Effect of Fe-supplementation on the expression of host iron-responsive genes in Mtb-infected rabbits. Heat map of selected immune response gene expression in the blood (**A–D**) or lung (**E–H**) during acute (4 and 8 weeks; **A**,**B**,**E**,**F**) or chronic (12 and 16 weeks; **C**,**D**,**G**,**H**) stages of infection in rabbits treated with Fe. The data from Fe-supplemented animals was calibrated with the corresponding levels in placebo-treated animals. Host house-keeping gene (*GAPDH)* expression was used to normalize target gene expression. Red color indicates up-regulation and blue color indicates down-regulation. The gradient in color indicates the trend towards up- or down-regulation.

3.4. Fe-Supplementation Perturbs Expression of Host Iron-Responsive Genes in the Lungs of Mtb-Infected Rabbits

The effect of Fe-supplementation on the expression of *HFE1*, *HFE2*, *HFE3* (*TFR2)*, *BMP6*, *HAMP*, *FPN1*, *NRAMP2*, *HMOX1*, *FTH1*, *NRF2*, and *LCN2* was assessed in rabbit lung homogenates. Expression of all the tested genes was significantly up-regulated in the lungs of Mtb-infected/placebo treated rabbits at 8 (acute) and 16 (chronic) weeks post-infection, compared to uninfected animals (Figure 3C,D). A similar trend in the expression of these genes was noted in Mtb-infected rabbit lungs at 4 (acute) and 12 (chronic) weeks post infection (Supplementary Figure S4). Fe-supplementation resulted in a significant down-regulation of *HFE3* (*TFR2*), *HAMP*, *FPN1*, and *LCN2* while *BMP6* and *FTH1* were significantly up-regulated, compared to the placebo group, at 4 weeks post-infection (Figure 4E). At 8 weeks post-infection, expression of *BMP6, HAMP, FPN1, FTH1, NRF2,* and *LCN2* was significantly up-regulated, while *HFE1* was significantly down-regulated by Fe-supplementation, compared to the placebo group (Figure 4F). Expression of *BMP6* and *FTH1* were up-regulated by Fe-supplementation at both, 4 and 8 weeks post-infection.

Among the tested Fe-responsive genes, expression of *HFE1, BMP6,* and *HMOX1* at 12 weeks, and *HAMP* at 16 weeks post-infection, was significantly up-regulated. At these later time points, expression of *HFE1, HFE3 (TFR2),* and *FPN1* was significantly down-regulated by Fe-treatment, compared to the placebo group (Figure 4G,H). Taken together, these results indicate that Mtb infection induces the expression of host Fe-responsive genes in rabbit lungs and that a subset of these genes were differentially regulated by Fe-supplementation during acute and chronic stages of infection.

3.5. *Fe-Supplementation Affects Systemic Expression of Host Immune Response Genes in Mtb-Infected Rabbits*

To determine the association between host iron- and immune-response gene expression at systemic level, we measured the expression of proinflammatory (*IFNG, TNFA, IL1B, NOS2* and *IL6*) and antiinflammatory (*IL10, SMAD6,* and *SMAD7*) genes in the blood of Mtb-infected and placebo- or Fe-supplemented rabbits. While proinflammatory cytokines are involved in mounting a protective immunity, anti-inflammatory molecules are involved in wound healing and dampening of inflammatory response [10,19]. SMAD6 and SMAD7 are involved in HAMP expression through BMP signaling [23] (Supplementary Table S2). In the blood and lung homogenates of Mtb-infected rabbits, the level of expression of all tested genes was significantly up-regulated, at 8 (acute) and 16 (chronic) weeks post infection (Figure 5A–D). Similar trend in the expression of these genes was noted in the blood and lungs at 4 (acute) and 12 (chronic) weeks post infection (Supplementary Figure S5). In Mtb-infected/Fe-supplemented rabbits, expression of *IFNG, TNFA,* and *IL1B* was significantly down-regulated, whereas expression of *IL6* and *SMAD6* was up-regulated, compared to placebo-treated animals, at 4 weeks post-infection (Figure 6A). At 8 weeks post-infection, expression of *SMAD6* and *SMAD7* was significantly up-regulated in the Mtb-infected/Fe-supplemented animals, compared to placebo-treated rabbits (Figure 6B). Although expression of all the tested genes was up-regulated in the blood at 12 and 16 weeks post-infection, Fe-supplementation did not significantly alter the expression of any of the tested genes at these time points (Figure 6C,D).

These results suggest that expression of host immune-response genes was significantly up-regulated systemically during acute and chronic stages of Mtb infection. Fe-supplementation during acute, and not chronic stages of infection, reduced the expression of a subset of proinflammatory genes while up-regulating antiinflammatory genes at the systemic level.

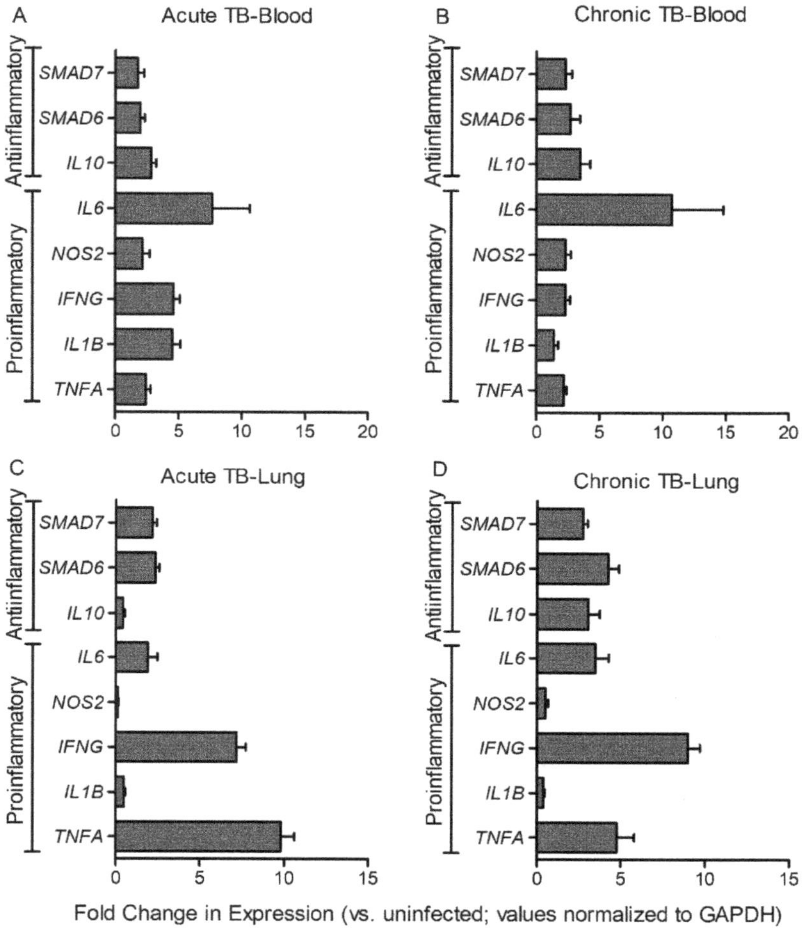

Figure 5. Expression of host pro- and anti-inflammatory response genes in Mtb-infected rabbits at 8 and 16 weeks post infection. Data shown are expression of target genes in the blood (**A**,**B**) or lung (**C**,**D**) during acute (8 weeks; **A**,**C**) or chronic (16 weeks; **B**,**D**) stages of infection. The gene expression levels in Mtb-infected animals was calibrated with the corresponding levels in uninfected rabbits. Host house-keeping gene (*GAPDH*) expression was used to normalize the level of target gene expression. Data was analyzed by one-way Anova with Tukey's multiple comparison test. Values plotted are mean +/− sd with $n = 4$ per group per time point. All tested genes were statistically significant ($p < 0.05$) in Figure 5A–D.

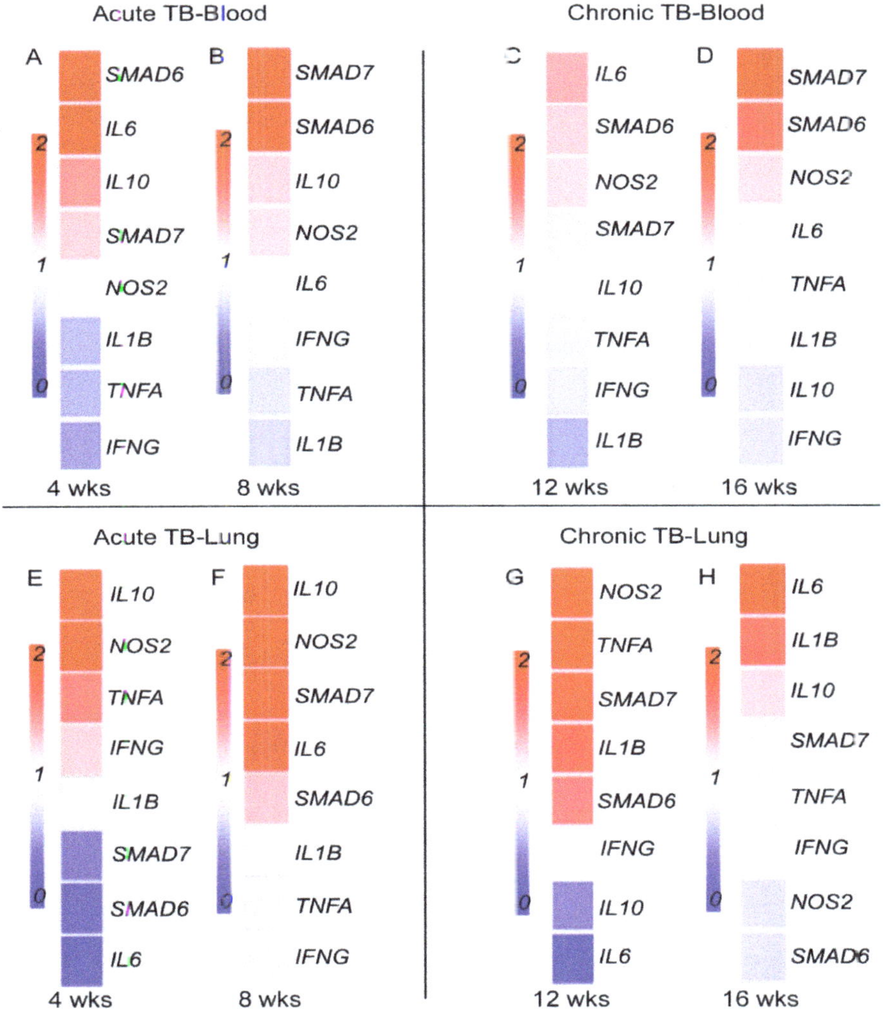

Figure 6. Effect of Fe-supplementation on the expression of host immune response genes in Mtb-infected rabbits. Heat map of selected immune response gene expression in the blood (**A–D**) or lung (**E–H**) during acute (4 and 8 weeks; **A**,**B**,**E**,**F**) or chronic (12 and 16 weeks; **C**,**D**,**G**,**H**) stages of infection in rabbits treated with Fe. The data from Fe-supplemented animals was calibrated with the corresponding levels in placebo-treated animals. Host house-keeping gene (*GAPDH*) expression was used to normalize target gene expression. Red color indicates up-regulation and blue color indicates down-regulation. The gradient in color indicates the trend towards up- or down-regulation

3.6. Fe-Supplementation Perturbs Expression of Host Immune Response Genes in the Lungs of Mtb-Infected Rabbits

Expression of *IFNG*, *TNFA*, *IL1B*, *NOS2*, *IL6*, *IL10*, *NOS2*, *SMAD6*, and *SMAD7* was measured at 4 and 8 (acute) or 12 and 16 (chronic) weeks post infection, using total RNA isolated from lung homogenates of rabbits with placebo- or Fe-supplementation (Figure 6E–H). Compared to uninfected, the level of expression of all tested genes were significantly up-regulated in the lungs during acute and chronic stages of Mtb infection, irrespective of treatment status. In Mtb-infected/Fe- supplemented rabbits, expression of *SMAD6* and *SMAD7* was significantly down-regulated at 4 weeks, compared to the placebo-treated animals (Figure 6E). At 8 weeks post-infection, expression of *IL10, NOS2,* and *SMAD7* was significantly up-regulated in the Fe-supplemented animals (Figure 6F).

Although the tested host immune response genes were highly up-regulated in the lungs at 12 and 16 weeks post-infection (Figure 5 D and Supplementary Figure S5), no significant difference in the expression level was observed for any of the tested genes between Fe-supplemented and placebo-treated rabbit lungs at these time points (Figure 6G,H).

These results suggest that Mtb infection up-regulates the expression of the tested host immune response genes at the site of infection, and a subset of these genes were differentially regulated by Fe-supplementation during acute infection.

3.7. Fe-Supplementation Did Not Affect the Lung Bacillary Load in Mtb-Infected Rabbits

To test whether Fe-supplementation affects the progression of initial infection to active disease (acute phase), Mtb CFU was determined at 4 and 8 weeks in the lung homogenates of rabbits with/without Fe-supplementation (Figure 7A). In Mtb-infected/placebo-treated animals, the lung bacterial CFU increased significantly, reaching a peak load at 4 weeks (3$\log_{10}$ at T = 0 to 4.5 $\log_{10}$ CFU at 4 weeks). A gradual reduction to 3.5 $\log_{10}$ CFU was seen at 8 weeks post-infection. Compared to the placebo group, Fe-supplementation from day 1 onwards until 8 weeks post-infection, did not significantly affect the bacterial CFU at both 4 and 8 weeks post-infection (Figure 7A). At these time points, no significant difference in body weight was seen between Mtb-infected/Fe-supplemented and placebo groups (Supplementary Figure S6).

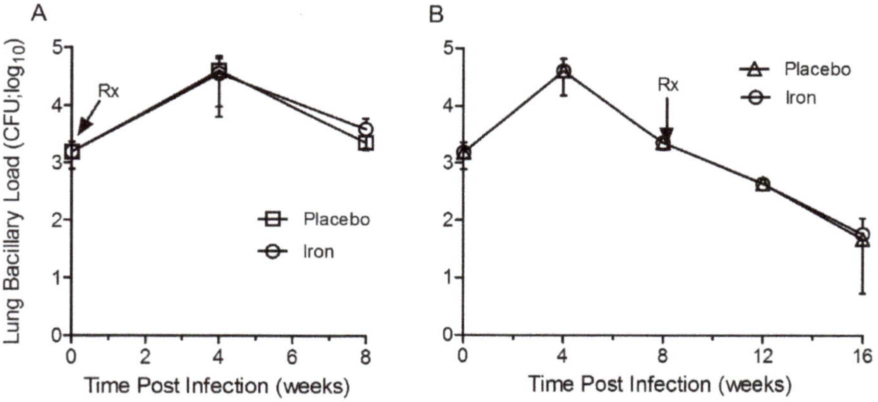

Figure 7. Effect of Fe-supplementation on the lung bacillary load of Mtb-infected rabbits. Number of bacterial colony-forming units (CFU) was determined in the lungs of rabbits supplemented with Fe or placebo during acute (**A**) or chronic (**B**) stages of Mtb infection. Data was analyzed by one-way Anova with Tukey's multiple comparison test. Values plotted are mean +/− sd with $n = 5$ per group per time point. Rx denotes treatment start time (day-1 or 8 weeks post-infection). The treatment was continued for 8 weeks in both acute and chronic infection groups.

To test whether Fe-supplementation affects the pulmonary bacterial burden during the chronic stage of infection, groups of Mtb-infected rabbits were supplemented with Fe or placebo for 8 weeks, starting at 8-week post-infection (until 16-week post-infection) (Figure 7B). In the placebo-treated animals, the lung bacillary load increased significantly with a peak burden of ~4.5 $\log_{10}$ CFU at four weeks, followed by a gradual decrease in CFU until 16 weeks post-infection, when the mean bacterial CFU was ~1.5 $\log_{10}$. Fe-supplementation of Mtb-infected rabbits from 8–16 weeks post-infection did not significantly affect the net cultivable bacillary load in the lungs, compared to the placebo group.

(Figure 7B). No significant difference in body weight was seen between Mtb-infected/Fe-supplemented and placebo-treated rabbits at these time points (Supplementary Figure S6).

These results suggest that Fe-supplementation during acute or chronic stages of Mtb infection did not significantly affect the bacterial burden in the lungs of infected rabbits.

3.8. Fe-Supplementation Did Not Affect Disease Pathology in Rabbit Lungs Infected with Mtb

To determine the effect of Fe-supplementation on TB immunopathology, we performed histologic staining of lung, liver, and spleen tissues with H&E, Perls' Fe-staining, or AFB staining to visualize immune cells, Fe deposition, or Mtb, respectively (Figure 8).

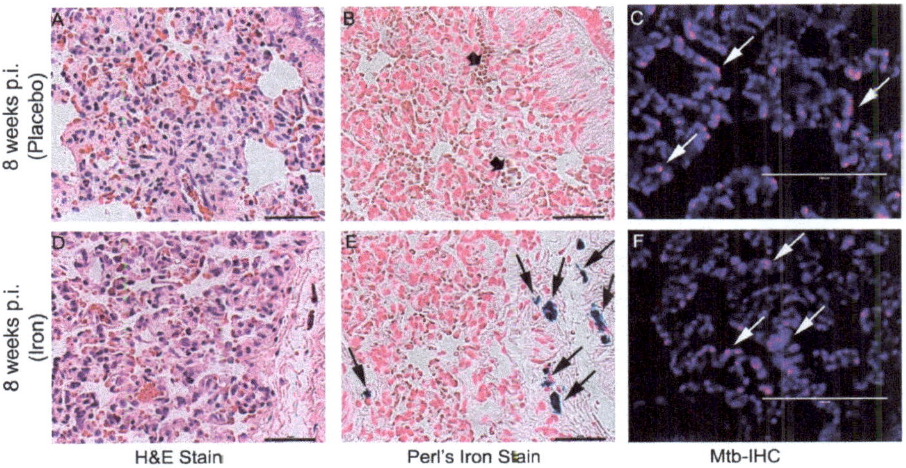

Figure 8. Effect of Fe-supplementation on rabbit lung pathology at 8 weeks post Mtb infection. Histopathology of rabbit lungs infected with Mtb CDC1551 at 8 weeks post infection with (**A–C**) or without (**D–F**) Fe-supplementation showing disease pathology (H&E stain; **A**,**D**), iron deposition (Perls' iron stain; **B**,**E**) and Mtb (by immunohistochemistry; **C**,**F**). Dark arrows in (**E**) show cellular iron deposition (blue color). White arrows in (**C**,**F**) show Mtb (purple color). The scale bar for all the images is 50 µm. Sections were photographed at 400× (**A**,**B**,**D**,**E**) or 600× (**C**,**F**) of original magnification.

Analysis of the H&E-stained lung sections of Mtb-infected/placebo-treated rabbits at 4 and 8 weeks showed both histopathologic and disease manifestations, as described previously (Figure 8 and Supplementary Figure S7) [19]. No clearly visible, macroscopic sub-pleural granuloma was seen in any of the infected rabbits at these time points. However, microscopic granulomas with increased immune cell infiltration, including neutrophils and monocytes, were prominent at 4 and 8 weeks post-infection, although no central necrosis or caseation was noticed in the granulomas. The granulomas in Mtb-infected/Fe-supplemented rabbit lungs at 4 and 8 weeks were very similar in size, cellular architecture, and maturation state compared to Mtb-infected/placebo-treated animals (Figure 8 and Supplementary Figure S7).

Fe deposition in lung granulomas was evaluated by Perls' staining method. Fe deposition was observed in the histiocytes of lung lesions only in Fe-supplemented and not in placebo-treated rabbits at 4 (Supplementary Figure S7) and 8 weeks post-infection/treatment (Figure 8B,E). At these time points, the presence of Mtb was observed by immunohistochemistry using anti-Mtb antibody in the lung lesions of both Fe-supplemented and placebo-treated, rabbits (Figure 8C, F and Supplementary Figure S7).

Histologic analysis of the H&E-stained lung sections of Mtb-infected/placebo-treated rabbits during chronic stages of infection (i.e, 12 and 16 weeks post infection) showed similar histopathology and disease manifestations, as described previously (Figure 9 and Supplementary Figure S8) [19]. No macroscopic sub-pleural lesions were observed in Mtb-infected rabbit lungs at these time points. However, a small number of microscopic cellular aggregations, reminiscent of resorbing lesions, with fewer immune cells present, were prominent at 12 weeks post-infection (Supplementary Figure S8). By 16 weeks, the lungs had a small number of cellular aggregates within the otherwise normal-looking lung parenchyma, consistent with our previous report (Figure 9) [19]. The lung lesions of Fe-supplemented rabbits at 12 and 16 weeks post-infection were very similar in size and cellular infiltration to the placebo-treated animals at the same time points (Figure 9 and Supplementary Figure S8). Perls' staining method revealed Fe deposition in the lung cells at 12 and 16 weeks post-infection/treatment only in the Fe-supplemented, and not in placebo-treated, rabbits (Figure 9B,E and Supplementary Figure S8). At these time points, Mtb was present in the lung sections of Fe-supplemented and placebo-treated animals (Figure 9C,F and Supplementary Figure S8).

These observations show that although elevated Fe was present in the lungs of Mtb-infected/Fe-supplemented rabbits at both acute and chronic stages of infection, the course of infection and disease pathology were not significantly altered by the supplementation.

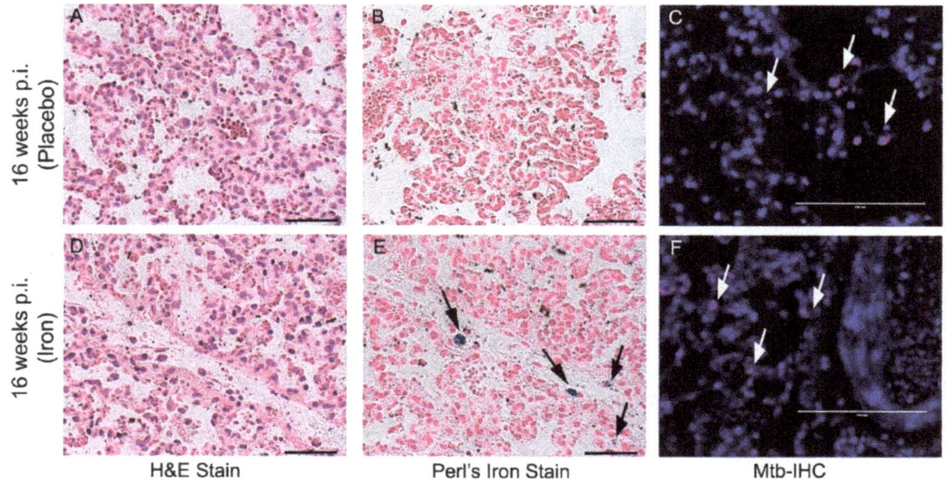

Figure 9. Effect of Fe-supplementation on rabbit lung pathology at 16-weeks post Mtb infection. Histopathology of rabbit lungs infected with Mtb CDC1551 at 16 weeks post infection with (**A–C**) or without (**D–F**) Fe-supplementation showing disease pathology (H&E stain; **A**,**D**), iron deposition (Perls' iron stain; **B**,**E**) and Mtb (by immunohistochemistry; **C**,**F**). Dark arrows in (**E**) show cellular iron deposition (blue color). White arrows in (**C**,**F**) show Mtb (purple color). The scale bar for all the images is 50 µm. Sections were photographed at 400× (**A**,**B**,**D**,**E**) or 600× (**C**,**F**) of original magnification.

4. Discussion

Here, we have determined the effect of Fe-supplementation on the outcome of non-progressive, pulmonary Mtb CDC1551 infection in a rabbit model. We show that Fe-supplementation alters the level of TIBC, total Fe, and %Tf saturation, and expression of host iron- and immune-response genes at the systemic and local levels during Mtb infection. Histologic examination of rabbit lungs revealed accumulation of Fe in Mtb-infected/Fe-supplemented rabbits without significant change in lung bacterial burden or disease pathology, compared to placebo-treated rabbits.

We observed significant up-regulation of Fe-responsive host gene expression in the lungs and in the blood of rabbits during acute and chronic stages of Mtb infection. Our findings are consistent with, and supported by previous reports in vitro and in mice, guinea pig, and non-human primate models of pulmonary TB [33–36]. These studies have shown that during Mtb infection, Fe-responsive host genes, such as *FPN1*, *FTH1*, *LCN2*, *HMOX1*, and *HAMP1*, were differentially regulated as a host-protective response against infecting bacteria. Some of these genes, such as *HAMP1* and *BMP6*, were also reported to be differentially expressed between lungs and liver of mycobacteria-infected mice [37]. Although gene expression in blood was not reported in these studies, we observed a differential expression pattern of host iron-responsive genes between blood and lung cells in Mtb-infected rabbits. We also observed differential TIBC, total Fe, and %Tf saturation levels between blood and lung homogenate of Mtb-infected rabbits. The association between Fe parameters and host iron-responsive gene expression in blood and lungs, and the relevance of this association to Mtb pathogenesis, requires further investigation.

We observed that Fe-supplementation did not significantly alter the expression of host iron-responsive genes that were up-regulated by Mtb infection in rabbit blood and lungs. The up-regulation of these genes, including *HAMP1*, *BMP6*, and *FTH1* in the lungs of Fe-supplemented rabbits, is likely part of the host's response to increased Fe accumulation. This is supported by histologic examination, which revealed accumulation of Fe in the lungs of Mtb-infected/Fe-supplemented rabbits. The induction of *HAMP* suggests down-regulation of Fe absorption and retention in macrophages; increased *HAMP* expression through induction of BMP pathway has been reported in human immune cells exposed to high Fe and in a mouse model of Mtb infection supplemented with Fe [38,39] Taken together, these results suggest that: (a) Mtb infection induces the expression of host iron-responsive genes in rabbits—a subset of these genes were differentially regulated by Fe-supplementation and/or by infection stage (acute versus chronic); and (b), host iron-responsive genes are differently regulated in blood and lungs of Mtb-infected rabbits.

During the course of Mtb infection, expression levels of many host immune-response genes were up-regulated both systemically (blood) and locally (lungs) in untreated rabbits. Most of these genes, including *TNFA*, *IL1B*, *IFNG*, and *IL10*, were up-regulated during acute phase of infection and their expression gradually declined, following immunological control the infection, at later stages in *Mtb*-infected rabbits. These findings are consistent with our earlier reports on humans, and in various animal models, that showed elevated expression of pro- and anti- inflammatory genes during Mtb infection [19,40]. Fe-supplementation of Mtb-infected rabbits reduced the expression of these host immune response genes systemically, particularly during acute phase of infection, but not in the lungs. However, expression of *IL6*, *IL10*, *SMAD6*, and *SMAD7* were up-regulated both in the lungs and in blood of Mtb-infected/Fe-supplemented rabbits. This is consistent with previous findings showing up-regulation of antiinflammatory cytokine genes, such as *IL10*, and down-regulation of the proinflammatory Th1 response during Fe-supplementation [10]. Similarly, reduction of *TNFA*, *IFNG*, and *IL1B* expression, with concomitant induction of *IL10* expression, has been reported previously in Fe-treated *M. bovis* BCG-infected mice, in Mtb-infected murine cell line J774, and in human mononuclear phagocytes [37,41,42]. In addition, up-regulation of IL-6, as a consequence of an acute response due to proinflammatory stimuli, has been shown to induce HAMP production in THP-1 cells, human macrophages, and in a murine model [38,43,44]. Taken together, these findings are consistent with our observations, and highlight the influence of Fe in the host immune response to Mtb infection.

We observed that perturbation of either the systemic or the local expression of host immune- and iron-responsive genes during acute and chronic stages of Mtb infection did not affect the outcome of infection. The organ bacillary loads, as well as disease pathology, were similar between Fe-supplemented and placebo-treated Mtb-infected rabbits. This finding is consistent with the data from recent studies in murine models of pulmonary TB with Fe-supplementation which showed that neither Fe-supplementation nor Fe deficiency significantly altered TB susceptibility and that Mtb load in infected organs was similar between Fe-supplemented/Fe-deficient and control mice [15,34,37]. One possibility for this observation is that although elevated Fe accumulation was noted in Mtb-infected/Fe-supplemented hosts, the Fe may not be available to infecting bacteria to enhance their growth and/or to cause exacerbated disease [45]. Consistently, %Tf saturation levels were significantly elevated in Fe-supplemented/Mtb-infected rabbits. Transferrin can sequester excess

Fe in the host, thus restricting its availability to Mtb. In contrast, another study showed increased Mtb growth upon Fe overload in a mouse model of pulmonary infection [46]. Similarly, exacerbated disease burden was reported in a guinea pig model of Mtb infection with Fe overload [47]. One reason for the discrepancy between these reports and our study on Mtb infection and Fe overload is that most of the animal and human studies on Fe-supplementation were conducted in an active TB condition, marked with progressive Mtb infection/disease, exacerbated inflammation, and bacillary loads in affected tissues [46,47]. In contrast, the rabbit model of LTBI used in this study develops non-progressive infection, with an initial protracted period of bacterial growth and a protracted granulomatous response, followed by a gradual reduction in bacillary load and disease pathology, both of which are efficiently controlled over time as the rabbits develop latency [19]. Therefore, the impact of Fe-supplementation is likely affected by the model system, the nature of infecting Mtb strain, and the subsequent host response.

The relationship between host Fe status and the outcome of Mtb infection is complex and not yet fully understood [10,11,34,37,38]. Patients with active TB, and some animal models of Mtb infection, have been shown to develop anemia, resulting from both Fe deficiency and/or chronic inflammation [48–50]. However, the association between host susceptibility to mycobacterial infection and host Fe status remains unclear [34,48–57]. A recent study showed that almost half of all patients with active pulmonary TB had anemia due to Fe deficiency and/or inflammation, at the time of diagnosis, and were recommended Fe supplementation as part of their therapy [57]. However, the association between anemia and LTBI remains unclear, although anemia is considered as a major risk factor for disease reactivation among these individuals [55–57]. Moreover, it has been reported that Fe-supplementation is beneficial to the host in situations where Fe deficiency co-exists with active disease showing exacerbated disease pathology [58]. In other conditions, where disease-associated inflammation predominates without concurrent Fe deficiency, the overall benefit of Fe-supplementation on the host is either negligible or insignificant [49]. In support of this notion, a study reported that Fe-supplementation to 131 pulmonary TB patients did not influence Mtb growth or disease severity, compared to the placebo group [49]. Similarly, reports on household contacts and LTBI individuals showed no significant correlation between Fe parameters and progression of disease [56]. Thus, in chronic, non-progressive Mtb infections, such as our study reported here, in which there was no exacerbated inflammation and associated disease pathology, Fe-supplementation may not have a significant influence on the outcome of Mtb infection.

In conclusion, our findings did not show a causal role for Fe supplementation and reactivation of LTBI in terms of bacterial burden and tissue pathology, although we observed changes in host gene expression associated with Fe-homeostasis and host immunity in Fe-supplemented animals,. However, results from this study, which links moderate Fe-supplementation with the host immunity during latent Mtb infection, should facilitate future investigations to determine the components of immune system that modulate Fe-dependent responses and to identify Mtb factors that perturb host Fe homeostasis.

Supplementary Materials: The following are available online at http://www.mdpi.com/2077-0383/8/8/1155/s1, Table S1: Description of primers used in this study; Table S2: Description of Fe and immune genes tested in this study; Figure S1: Scheme for rabbit Mtb infection and Fe-treatment; Figure S2: Effect of Fe-treatment on the hematocrit (top) and hemoglobin (bottom) content of rabbits with acute (A) or chronic (B) Mtb infection; Figure S3: Effect of Fe-supplementation on the lung iron parameters of rabbits at 4 and 12 weeks post infection; Figure S4: Expression of iron-responsive genes in Mtb-infected rabbits at 4 and 12 weeks post infection; Figure S5: Expression of host pro- and anti-inflammatory response genes in Mtb-infected rabbits at 4 and 12 weeks post infection; Figure S6: Body weight of Mtb-infected rabbits with or without Fe-treatment; Figure S7: Effect of Fe supplementation on rabbit lung pathology at 4 weeks post Mtb infection; Figure S8: Effect of Fe supplementation on rabbit lung pathology at 12 weeks post Mtb infection.

Author Contributions: S.S. and G.M.R. conceived and designed the study. A.K., P.S., and S.S. performed the experiments and analyzed the data. A.K., P.S., G.M.R., and S.S. interpreted the data. A.K., P.S., and S.S. wrote the manuscript. All authors have read, reviewed and edited the manuscript and agreed for submission to this journal.

Funding: This study was funded by NIH/NIAID grant to S.S. (AI119619). The funding agency has no role in the study design, data collection, analysis, interpretation or writing the manuscript.

Conflicts of Interest: The authors declare no conflict of interest.

References

1. Dye, C.; Williams, B.G. The population dynamics and control of tuberculosis. *Science* **2010**, *328*, 856–861. [CrossRef] [PubMed]
2. Behr, M.A.; Edelstein, P.H.; Ramakrishnan, L. Revisiting the timetable of tuberculosis. *BMJ* **2018**, *362*, k2738. [CrossRef] [PubMed]
3. O'Garra, A.; Redford, P.S.; McNab, F.W.; Bloom, C.I.; Wilkinson, R.J.; Berry, M.P. The immune response in tuberculosis. *Annu. Rev. Immunol.* **2013**, *31*, 475–527. [CrossRef] [PubMed]
4. Ernst, J.D. The immunological life cycle of tuberculosis. *Nat. Rev. Immunol.* **2012**, *12*, 581–591. [CrossRef] [PubMed]
5. Ganz, T.; Nemeth, E. Iron homeostasis in host defence and inflammation. *Nat. Rev. Immunol.* **2015**, *15*, 500–510. [CrossRef] [PubMed]
6. Michels, K.R.; Zhang, Z.; Bettina, A.M.; Cagnina, R.E.; Stefanova, D.; Burdick, M.D.; Vaulont, S.; Nemeth, E.; Ganz, T.; Mehrad, B. Hepcidin-mediated iron sequestration protects against bacterial dissemination during pneumonia. *JCI Insight* **2017**, *2*, e92002. [CrossRef] [PubMed]
7. Stefanova, D.; Raychev, A.; Arezes, J.; Ruchala, P.; Gabayan, V.; Skurnik, M.; Dillon, B.J.; Horwitz, M.A.; Ganz, T.; Bulut, Y.; et al. Endogenous hepcidin and its agonist mediate resistance to selected infections by clearing non-transferrin-bound iron. *Blood* **2017**, *130*, 245–257. [CrossRef]
8. Chao, A.; Sieminski, P.J.; Owens, C.P.; Goulding, C.W. Iron acquisition in Mycobacterium tuberculosis. *Chem. Rev.* **2019**, *119*, 1193–1220. [CrossRef] [PubMed]
9. Clemens, D.L.; Horwitz, M.A. The Mycobacterium tuberculosis phagosome interacts with early endosomes and is accessible to exogenously administered transferrin. *J. Exp. Med.* **1996**, *184*, 1349–1355. [CrossRef]
10. Boelaert, J.R.; Vandecasteele, S.J.; Appelberg, R.; Gordeuk, V.R. The effect of the host's iron status on tuberculosis. *J. Infect. Dis.* **2007**, *195*, 1745–1753. [CrossRef]
11. Ratledge, C. Iron, mycobacteria and tuberculosis. *Tuberculosis* **2004**, *84*, 110–130. [CrossRef] [PubMed]
12. Rodriguez, G.M. Control of iron metabolism in Mycobacterium tuberculosis. *Trends Microbiol.* **2006**, *14*, 320–327. [CrossRef] [PubMed]
13. Kurthkoti, K.; Amin, H.; Marakalala, M.J.; Ghanny, S.; Subbian, S.; Sakatos, A.; Livny, J.; Fortune, S.M.; Berney, M.; Rodriguez, G.M. The capacity of *Mycobacterium tuberculosis* to survive iron starvation might enable it to persist in iron-deprived microenvironments of human granulomas. *MBio* **2017**, *8*, e01092-17. [CrossRef] [PubMed]
14. Tufariello, J.M.; Chapman, J.R.; Kerantzas, C.A.; Wong, K.W.; Vilchèze, C.; Jones, C.M.; Cole, L.E.; Tinaztepe, E.; Thompson, V.; Fenyö, D.; et al. Separable roles for Mycobacterium tuberculosis ESX-3 effectors in iron acquisition and virulence. *Proc. Natl. Acad. Sci. USA* **2016**, *113*, E348–E357. [CrossRef] [PubMed]
15. Madigan, C.A.; Martinot, A.J.; Wei, J.R.; Madduri, A.; Cheng, T.Y.; Young, D.C.; Layre, E.; Murry, J.P.; Rubin, E.J.; Moody, D.B. Lipidomic analysis links mycobactin synthase K to iron uptake and virulence in M. tuberculosis. *PLoS Pathog.* **2015**, *11*, e1004792. [CrossRef] [PubMed]
16. Kochan, I. The role of iron in bacterial infections, with special consideration of host-tubercle bacillus interaction; *Current Topics in Microbiology and Immunology*; Springer: Berlin, Heidelberg, Germany, 1973; pp. 1–30.
17. Murray, M.J.; Murray, A.B.; Murray, M.B.; Murray, C.J. The adverse effect of iron repletion on the course of certain infections. *Br. Med. J.* **1978**, *2*, 1113–1115. [CrossRef]
18. Lounis, N.; Maslo, C.; Truffot-Pernot, C.; Grosset, J.; Boelaert, R.J. Impact of iron loading on the activity of isoniazid or ethambutol in the treatment of murine tuberculosis. *Int. J. Tuberc. Lung Dis.* **2003**, *7*, 575–579. [PubMed]
19. Subbian, S.; Tsenova, L.; O'Brien, P.; Yang, G.; Kushner, N.L.; Parsons, S.; Peixoto, B.; Fallows, D.; Kaplan, G. Spontaneous latency in a rabbit model of pulmonary tuberculosis. *Am. J. Pathol.* **2012**, *181*, 1711–1724. [CrossRef]

20. Rashtchizadeh, N.; Ettehad, S.; DiSilvestro, R.A.; Mahdavi, R. Antiatherogenic effects of zinc are associated with copper in iron-overloaded hypercholesterolemic rabbits. *Nutr. Res.* **2008**, *28*, 98–105. [CrossRef]
21. Deschemin, J.C.; Mathieu, J.R.R.; Zumerle, S.; Peyssonnaux, C.; Vaulont, S. Pulmonary iron homeostasis in hepcidin knockout mice. *Front. Physiol.* **2017**, *8*, 804. [CrossRef]
22. Subbian, S.; Eugenin, E.; Kaplan, G. Detection of Mycobacterium tuberculosis in latently infected lungs by immunohistochemistry and confocal microscopy. *J. Med. Microbiol.* **2014**, *63*, 1432–1435. [CrossRef] [PubMed]
23. Avecilla, S.T.; Marionneaux, S.M.; Leiva, T.D.; Tonon, J.A.; Chan, V.T.; Moung, C.; Meagher, R.C.; Maslak, P. Comparison of manual hematocrit determinations versus automated methods for hematopoietic progenitor cell apheresis products. *Transfusion* **2016**, *56*, 528–532. [CrossRef] [PubMed]
24. Subbian, S.; O'Brien, P.; Kushner, N.L.; Yang, G.; Tsenova, L.; Peixoto, B.; Bandyopadhyay, N.; Bader, J.S.; Karakousis, P.C.; Fallows, D.; et al. Molecular immunologic correlates of spontaneous latency in a rabbit model of pulmonary tuberculosis. *Cell Commun. Signal.* **2013**, *11*, 16. [CrossRef] [PubMed]
25. Rodriguez Martinez, A.; Niemela, O.; Parkkila, S. Hepatic and extrahepatic expression of the new iron regulatory protein hemojuvelin. *Haematologica* **2004**, *89*, 1441–1445. [PubMed]
26. D'Alessio, F.; Hentze, M.W.; Muckenthaler, M.U. The hemochromatosis proteins HFE, TfR2, and HJV form a membrane-associated protein complex for hepcidin regulation. *J. Hepatol.* **2012**, *57*, 1052–1060. [CrossRef]
27. Babitt, J.L.; Huang, F.W.; Wrighting, D.M.; Xia, Y.; Sidis, Y.; Samad, T.A.; Campagna, J.A.; Chung, R.T.; Schneyer, A.L.; Woolf, C.J.; et al. Bone morphogenetic protein signaling by hemojuvelin regulates hepcidin expression. *Nat. Genet.* **2006**, *38*, 531–539. [CrossRef] [PubMed]
28. Sabelli, M.; Montosi, G.; Garuti, C.; Caleffi, A.; Oliveto, S.; Biffo, S.; Pietrangelo, A. Human macrophage ferroportin biology and the basis for the ferroportin disease. *Hepatology* **2017**, *65*, 1512–1525. [CrossRef]
29. Arosio, P.; Elia, L.; Poli, M. Ferritin, cellular iron storage and regulation. *IUBMB Life* **2017**, *69*, 414–422. [CrossRef]
30. El-Rifaie, A.A.; Sabry, D.; Doss, R.W.; Kamal, M.A.; Abd El Hassib, D.M. Heme oxygenase and iron status in exosomes of psoriasis patients. *Arch. Dermatol. Res.* **2018**, *310*, 651–656. [CrossRef]
31. Kasai, S.; Mimura, J.; Ozaki, T.; Itoh, K. Emerging regulatory role of Nrf2 in iron, heme, and hemoglobin metabolism in physiology and disease. *Front. Vet. Sci.* **2018**, *5*, 242. [CrossRef]
32. Flo, T.H.; Smith, K.D.; Sato, S.; Rodriguez, D.J.; Holmes, M.A.; Strong, R.K.; Akira, S.; Aderem, A. Lipocalin 2 mediates an innate immune response to bacterial infection by sequestrating iron. *Nature* **2004**, *432*, 917–921. [CrossRef] [PubMed]
33. Wareham, A.S.; Tree, J.A.; Marsh, P.D.; Butcher, P.D.; Dennis, M.; Sharpe, S.A. Evidence for a role for interleukin-17, Th17 cells and iron homeostasis in protective immunity against tuberculosis in cynomolgus macaques. *PLoS ONE* **2014**, *9*, e88149. [CrossRef] [PubMed]
34. Harrington-Kandt, R.; Stylianou, E.; Eddowes, L.A.; Lim, P.J.; Stockdale, L.; Pinpathomrat, N.; Bull, N.; Pasricha, J.; Ulaszewska, M.; Beglov, Y.; et al. Hepcidin deficiency and iron deficiency do not alter tuberculosis susceptibility in a murine M.tb infection model. *PLoS ONE* **2018**, *13*, e0191038.
35. McDermid, J.M.; Prentice, A.M. Iron and infection: Effects of host iron status and the iron-regulatory genes haptoglobin and NRAMP1 (SLC11A1) on host-pathogen interactions in tuberculosis and HIV. *Clin. Sci.* **2006**, *110*, 503–524. [CrossRef] [PubMed]
36. Thom, R.E.; Elmore, M.J.; Williams, A.; Andrews, S.C.; Drobniewski, F.; Marsh, P.D.; Tree, J.A. The expression of ferritin, lactoferrin, transferrin receptor and solute carrier family 11A1 in the host response to BCG-vaccination and Mycobacterium tuberculosis challenge. *Vaccine* **2012**, *30*, 3159–3168. [CrossRef] [PubMed]
37. Agoro, R.; Benmerzoug, S.; Rose, S.; Bouyer, M.; Gozzelino, R.; Garcia, I.; Ryffel, B.; Quesniaux, V.F.J.; Mura, C. An iron-rich diet decreases the Mycobacterial burden and correlates with hepcidin upregulation, lower levels of proinflammatory mediators, and increased T-cell recruitment in a model of Mycobacterium bovis Bacille Calmette-Guerin infection. *J. Infect. Dis.* **2017**, *216*, 907–918. [CrossRef] [PubMed]
38. Abreu, R.; Quinn, F.; Giri, P.K. Role of the hepcidin-ferroportin axis in pathogen-mediated intracellular iron sequestration in human phagocytic cells. *Blood Adv.* **2018**, *2*, 1089–1100. [CrossRef] [PubMed]
39. Paesano, R.; Natalizi, T.; Berlutti, F.; Valenti, P. Body iron delocalization: The serious drawback in iron disorders in both developing and developed countries. *Pathog. Glob. Health* **2012**, *106*, 200–216. [CrossRef]

40. Domingo-Gonzalez, R.; Prince, O.; Cooper, A.; Khader, S.A. Cytokines and chemokines in Mycobacterium tuberculosis infection. *Microbiol. Spectr.* **2016**, *4*. [CrossRef]
41. Serafin-Lopez, J.; Chacon-Salinas, R.; Munoz-Cruz, S.; Enciso-Moreno, J.A.; Estrada-Parra, S.A.; Estrada-Garcia, I. The effect of iron on the expression of cytokines in macrophages infected with Mycobacterium tuberculosis. *Scand. J. Immunol.* **2004**, *60*, 329–337. [CrossRef]
42. Byrd, T.F. Tumor necrosis factor alpha (TNFalpha) promotes growth of virulent Mycobacterium tuberculosis in human monocytes iron-mediated growth suppression is correlated with decreased release of TNFalpha from iron-treated infected monocytes. *J. Clin. Invest.* **1997**, *99*, 2518–2529. [CrossRef] [PubMed]
43. Rodriguez, R.; Jung, C.L.; Gabayan, V.; Deng, J.C.; Ganz, T.; Nemeth, E.; Bulut, Y. Hepcidin induction by pathogens and pathogen-derived molecules is strongly dependent on interleukin-6. *Infect. Immun.* **2014**, *82*, 745–752. [CrossRef] [PubMed]
44. Armitage, A.E.; Eddowes, L.A.; Gileadi, U.; Cole, S.; Spottiswoode, N.; Selvakumar, T.A.; Ho, L.P.; Townsend, A.R.; Drakesmith, H. Hepcidin regulation by innate immune and infectious stimuli. *Blood* **2011**, *118*, 4129–4139. [CrossRef] [PubMed]
45. Nairz, M.; Dichtl, S.; Schroll, A.; Haschka, D.; Tymoszuk, P.; Theurl, I.; Weiss, G. Iron and innate antimicrobial immunity-depriving the pathogen, defending the host. *J. Trace Elem. Med. Biol.* **2018**, *48*, 118–133. [CrossRef] [PubMed]
46. Schaible, U.E.; Collins, H.L.; Priem, F.; Kaufmann, S.H. Correction of the iron overload defect in beta-2-microglobulin knockout mice by lactoferrin abolishes their increased susceptibility to tuberculosis. *J. Exp. Med.* **2002**, *196*, 1507–1513. [CrossRef]
47. Miles, A.A.; Khimji, P.L.; Maskell, J. The variable response of bacteria to excess ferric iron in host tissues. *J. Med. Microbiol.* **1979**, *12*, 17–28. [CrossRef]
48. Punnonen, K.; Irjala, K.; Rajamaki, A. Iron-deficiency anemia is associated with high concentrations of transferrin receptor in serum. *Clin. Chem.* **1994**, *40*, 774–776.
49. Das, B.S.; Devi, U.; Mohan Rao, C.; Srivastava, V.K.; Rath, P.K. Effect of iron supplementation on mild to moderate anaemia in pulmonary tuberculosis. *Br. J. Nutr.* **2003**, *90*, 541–550.
50. Fleck, A.; Myers, M.A. Diagnostic and prognostic significance of acute phase proteins. In *The Acute Phase Response to Injury and Infection*; Gordon, A.H., Ed.; Elsevier Science Publishers: Amsterdam, The Netherlands, 1985; pp. 249–271.
51. Gangaidzo, I.T.; Moyo, V.M.; Mvundura, E.; Aggrey, G.; Murphree, N.L.; Khumalo, H.; Saungweme, T.; Kasvosve, I.; Gomo, Z.A.; Rouault, T.; et al. Association of pulmonary tuberculosis with increased dietary iron. *J. Infect. Dis.* **2001**, *184*, 936–939. [CrossRef]
52. Tanner, R.; O'Shea, M.K.; White, A.D.; Muller, J.; Harrington-Kandt, R.; Matsumiya, M.; Dennis, M.J.; Parizotto, E.A.; Harris, S.; Stylianou, E.; et al. The influence of haemoglobin and iron on in vitro mycobacterial growth inhibition assays. *Sci. Rep.* **2017**, *7*, 43478. [CrossRef]
53. Iannotti, L.L.; Tielsch, J.M.; Black, M.M.; Black, R.E. Iron supplementation in early childhood: Health benefits and risks. *Am. J. Clin. Nutr.* **2006**, *84*, 1261–1276. [CrossRef] [PubMed]
54. Adetifa, I.; Okomo, U. Iron supplementation for reducing morbidity and mortality in children with HIV. *Cochrane Database Syst. Rev.* **2009**, *1*, CD006736. [CrossRef] [PubMed]
55. Minchella, P.A.; Donkor, S.; McDermid, J.M.; Sutherland, J.S. Iron homeostasis and progression to pulmonary tuberculosis disease among household contacts. *Tuberculosis* **2015**, *95*, 288–293. [CrossRef] [PubMed]
56. Takenami, I.; Loureiro, C.; Machado, A.; Emodi, K., Jr.; Riley, L.W.; Arruda, S. Blood cells and interferon-gamma levels correlation in latent tuberculosis infection. *ISRN Pulmonol.* **2013**, *2013*, 256148. [CrossRef] [PubMed]
57. Minchella, P.A.; Donkor, S.; Owolabi, O.; Sutherland, J.S.; McDermid, J.M. Complex anemia in tuberculosis: The need to consider causes and timing when designing interventions. *Clin. Infect. Dis.* **2015**, *60*, 764–772. [CrossRef]
58. Baer, A.N.; Dessypris, E.N.; Krantz, S.B. The pathogenesis of anemia in rheumatoid arthritis: A clinical and laboratory analysis. *Semin. Arthr. Rheum.* **1990**, *19*, 209–223. [CrossRef]

Journal of
Clinical Medicine

MDPI

Article

Use of Antiplatelet Agents and Survival of Tuberculosis Patients: A Population-Based Cohort Study

Meng-Rui Lee [1,2,3], Ming-Chia Lee [4,5], Chia-Hao Chang [1,2], Chia-Jung Liu [1,2], Lih-Yu Chang [1,2], Jun-Fu Zhang [2], Jann-Yuan Wang [2] and Chih-Hsin Lee [6,7,*]

1 Department of Internal Medicine, National Taiwan University Hospital, Hsin-Chu Branch, Hsinchu 30059, Taiwan
2 Department of Internal Medicine, National Taiwan University Hospital, Taipei 10002, Taiwan
3 Institute of Epidemiology and Preventive Medicine, College of Public Health, National Taiwan University, Taipei 10052, Taiwan
4 Department of Pharmacy, New Taipei City Hospital, New Taipei City 24141, Taiwan
5 School of Pharmacy, College of Pharmacy, Taipei Medical University, Taipei 11031, Taiwan
6 Department of Internal Medicine, School of Medicine, College of Medicine, Taipei Medical University, Taipei 11031, Taiwan
7 Pulmonary Research Center, Division of Pulmonary Medicine, Wan Fang Hospital, Taipei Medical University, Taipei 11696, Taiwan
* Correspondence: chleetw@tmu.edu.tw; Fax: +886-2866-21138

Received: 16 May 2019; Accepted: 24 June 2019; Published: 27 June 2019

check for updates

Abstract: While evidence is accumulating that platelets contribute to tissue destruction in tuberculosis (TB) disease, it is still not known whether antiplatelet agents are beneficial to TB patients. We performed this retrospective cohort study and identified incident TB cases in the Taiwan National Tuberculosis Registry from 2008 to 2014. These cases were further classified into antiplatelet users and non-users according to the use of antiplatelet agents prior to the TB diagnosis, and the cohorts were matched using propensity scores (PSs). The primary outcome was survival after a TB diagnosis. In total, 74,753 incident TB cases were recruited; 9497 (12.7%) were antiplatelet users, and 7764 (10.4%) were aspirin (ASA) users. A 1:1 PS-matched cohort with 8864 antiplatelet agent users and 8864 non-users was created. After PS matching, antiplatelet use remained associated with a longer survival (adjusted hazard ratio (HR): 0.91, 95% confidence interval (CI): 0.88–0.95, $p < 0.0001$). The risk of major bleeding was not elevated in antiplatelet users compared to non-users ($p = 0.604$). This study shows that use of antiplatelet agents has been associated with improved survival in TB patients. The immunomodulatory and anti-inflammatory effects of antiplatelet agents in TB disease warrant further investigation. Antiplatelets are promising as an adjunct anti-TB therapy.

Keywords: tuberculosis; antiplatelet; aspirin; immunomodulation; survival; Taiwan

1. Introduction

Tuberculosis (TB) remains an important global infectious disease with an estimated 10 million new cases and 1.6 million deaths in 2017 [1]. The World Health Organization (WHO) has set a goal of eradicating TB as a public health problem, aiming to achieve 50% and 90% reductions in the global TB incidence by 2025 and 2035, respectively [2]. TB remains an important infectious disease that causes significant morbidity and mortality [3]. Though considered a treatable infectious disease, TB treatment success rates and cure rates are still suboptimal [4]. The mortality rate from TB remains high, especially in elderly populations, which can exceed 20% in patients older than 65 years [5]. Besides point-of-care diagnostic methods, the discovery of newer anti-TB agents and a more-thorough understanding of its pathophysiology remain important clinical issues in combating this disease.

While the majority of TB studies regarding host immunity have focused on monocytes and lymphocytes, several recent basic science studies have highlighted the role of platelets as a novel participant in TB's pathogenesis [6]. It was first observed that platelet activation status was correlated with TB disease severity [7]. In active TB, a hypercoagulable state exists with an increased risk of thrombosis, which improves after treatment [8]. Furthermore, it is now recognized that platelets play a role in tissue destruction and pro-inflammation in TB disease [9]. In a recent study, active platelets were present at sites of pulmonary TB, and a co-culture with platelets decreased the intracellular killing of *Mycobacterium tuberculosis* and increased its replication [9]. Finally, another more recent study further provided in vivo experimental evidence that infection-induced platelet activation is a potential target for TB host directed therapy [10]. With accumulating evidence of platelet involvement in TB's pathophysiology, it was therefore intriguing to determine if the use of antiplatelet agents would be beneficial for TB patients and could serve as a potential adjunct anti-TB agent. However, no epidemiologic studies have examined this question.

We therefore initiated this population-based study to investigate the clinical impacts of antiplatelet agents on the survival of TB patients. Through the linkage of the Taiwan National Tuberculosis Registry (TNTR) database, the Taiwan National Health Insurance (NHI) database, and national mortality data, we recruited incident TB cases and evaluated whether antiplatelet agents were associated with better outcomes.

2. Experimental Section

2.1. Ethics Statement

The Institutional Review Board of Taipei Medical University approved the study (N201712019) and waved the need for informed consent because this retrospective study used encrypted data and presented no risk to participants.

2.2. Study Participants and Setting

This study was conducted by linking Taiwan NHI claims data, mortality data from the Department of Statistics, Ministry of Health, and Welfare, and the TNTR [11]. The TNTR was established by the Taiwan Centers for Disease Control (CDC) in 1996, and clinicians are obligated to report and register every TB patient in Taiwan in the TNTR [12,13]. In addition, the registry system includes information on TB characteristics, treatment courses, and clinical outcomes. Taiwan's NHI is a universal healthcare system that covers 96% of the residents of Taiwan (with a population of about 23 million) [14–17].

The inclusion criterion was incident TB cases who received anti-TB treatment identified from the TNTR between 2008 and 2014. Patients with multidrug-resistant TB (MDRTB), with incomplete data, or who were younger than 20 years were excluded.

2.3. Definitions and Data Collection

A diagnosis of TB and information regarding TB disease characteristics (smear positivity, culture positivity, and a cavitation on chest radiography) were ascertained from the TNTR. In Taiwan, a diagnosis of TB is made based on clinical symptoms, microbiological studies, radiographic findings, and response to anti-TB treatment [18]. Comorbidities and clinical characteristics of TB patients were extracted from the Taiwanese NHI claims database.

Antiplatelets were divided into aspirin (ASA, irreversible cyclooxygenase inhibitor) and non-ASA antiplatelets, including adenosine diphosphate (ADP) receptor inhibitors, phosphodiesterase inhibitors, glycoprotein IIB/IIIA inhibitors, and adenosine reuptake inhibitors (Supplementary Table S1). Protease-activated receptor (PAR)-1 antagonists and thromboxane receptor antagonists were not available in Taiwan during the study period and were therefore excluded from our study.

Users of each category of drugs were defined as using more than 90 defined daily doses (DDDs) of all drugs in the category within 180 days prior to the TB diagnosis. The calculation of DDD followed its definition by the WHO, which is the assumed average maintenance dose per day for a drug used for its main indication in adults [19].

We collected information (DDD) regarding the usage of statins, metformin, nonsteroidal anti-inflammatory drugs (NSAIDs), and corticosteroids within 180 days prior to the TB diagnosis.

The definition of comorbidities is summarized in Supplementary Table S2. Immunocompromised hosts were defined if they had either diabetes mellitus (DM), end-stage renal disease (ESRD), cancer, cirrhosis of the liver, steroid use, a transplant, or acquired immunodeficiency syndrome (AIDS).

2.4. Outcomes

The primary outcome was patient survival after a TB diagnosis. Secondary outcomes were mortality within 12 months after the diagnosis and major bleeding events after the diagnosis.

The definition of major bleeding was modified from the International Society on Thrombosis and Haemostasis (ISTH) definition, and modifications were made due to the limitations of the claims database [20,21]. The definition of major bleeding in our study was hospitalization after TB diagnosis due to either an intracranial hemorrhage or a gastrointestinal hemorrhage necessitating a transfusion. The International Classification of Diseases, Ninth Revision, Clinical Modification (ICD-9-CM) and ICD-10-CM codes for defining major bleeding are described in Supplementary Table S3. The date of hospitalization was the bleeding date.

All participants were followed up until end of the study period (31 December 2016).

2.5. Statistical Analysis

Proportions or means were used to describe the demographic, clinical, and radiographic characteristics of TB patients. Inter-group differences were analyzed using an independent-sample *t*-test for continuous variables and a chi-squared test for categorical variables. The propensity score (PS) for the probability of being administered an antiplatelet agent was derived using a logistic regression model including the potential confounders of age, sex, socioeconomic status, smear positivity, culture positivity, cavitation, comorbidities, the Charlson comorbidity index (CCI) [22], coronary artery disease, DM, ESRD, cancer, cirrhosis, ischemic stroke, chronic obstructive pulmonary disease, AIDS, and hypertension.

A Cox proportional hazard regression model and logistic regression were, respectively, used to analyze factors associated with patient survival and one-year mortality. PS-matching and PS-stratified analysis were both performed. A cumulative incidence function was used to analyze major bleeding events due to competing risks. Variables that remained significantly different after PS matching and accounting for concomitant medications, including statins, steroids, metformin, and NSAIDs, were further adjusted in the final model. We also performed survival and one-year mortality rate analyses before PS matching, adjusted with variables used in the PS derivation. Subgroup analyses were performed among immunocompromised TB patients, including DM, ESRD, psoriasis, cancer, and steroids users. All data analyses were performed using SAS version 9.4 (SAS Institute, Cary, NC, USA). $P < 0.05$ of a two-sided test was considered statistically significant.

3. Results

3.1. Patients Identification

The patient recruitment process is illustrated in Figure 1. In total, 74,753 participants were recruited for the study.

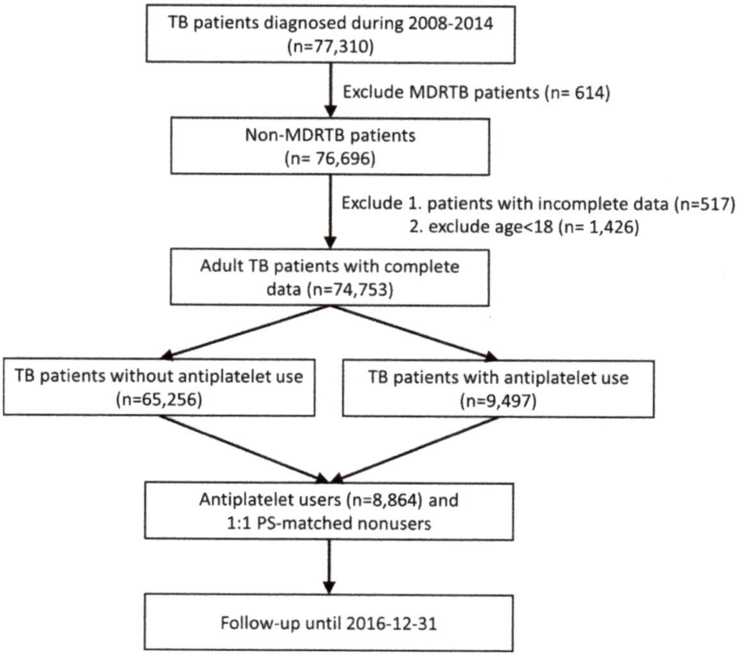

Figure 1. Flowchart of patient recruitment.

3.2. Demographic Data

The clinical characteristics of identified participants are described in Table 1. Among 74,753 participants, 12.7% (n = 9497) were antiplatelet users, and 10.4% (n = 7764) were aspirin users. The mean age was 63.6 years for all participants, and 69.8% were male. Compared to non-users, antiplatelet users were older, more likely to be male, and had higher CCI scores. Antiplatelet users were also more likely to have underlying comorbidities, except for AIDS, rheumatoid arthritis (RA), transplant, and pneumoconiosis. Antiplatelet users were also more likely to be culture positive, but they were less likely to be smear positive or have a cavitation on chest radiography. Antiplatelet users also took a higher cumulative dose of statins, NSAIDs, metformin, and corticosteroids (Supplementary Table S4).

After PS matching, 8864 antiplatelet users, including 7281 ASA users and 1704 non-ASA antiplatelet users, were matched with 8864 antiplatelet non-users. As for comorbidities, only the stroke incidence significantly differed between the two groups (26.9% versus 24.5%, $p < 0.001$). Antiplatelet users also took higher cumulative doses of statins and metformin (Supplementary Table S4).

For the subgroup analysis, a 1:1 comparison cohort was created for the ASA (7281 ASA users and 7281 matched antiplatelet non-users) and non-ASA antiplatelet (1704 non-ASA antiplatelet users and 1704 matched antiplatelet non-users) groups from the PS-matched population (Table 2). For the ASA subgroup analysis, only the stroke incidence significantly differed between ASA users and matched non-users (24.7% versus 22.6%, $p = 0.003$). ASA users also took higher cumulative doses of statins and metformin and a lower dose of corticosteroids compared to non-users (Supplementary Table S5). For the non-ASA antiplatelet subgroup analysis, only ESRD significantly differed between non-ASA antiplatelet users and matched non-users (10.9% versus 8.0%, $p = 0.005$) (Table 2). Non-ASA antiplatelet users also took a higher cumulative dose of statins compared to non-users (Supplementary Table S5).

Table 1. Clinical characteristics of tuberculosis patients with and those without antiplatelet use.

	Before PS matching						After PS matching					
	Overall (N = 74,753)	Antiplatelet users (N = 9497)			Non-users (n = 65,256)	p value *	Overall (n = 17,728)	Antiplatelet users (N = 8864)			Non-users (n = 8864)	p value *
		All (n = 9497)	ASA (n = 7764)	Non-ASA (n = 1855)				All (n = 8864)	ASA (n = 7281)	Non-ASA (n = 1704)		
Age (mean ± SD)	63.6 ± 18.5	76.1 ± 10.6	75.7 ± 10.7	77.5 ± 9.9	61.7 ± 19.7	< 0.001	75.9 ± 10.6	75.7 ± 10.6	75.4 ± 10.7	77.2 ± 9.9	76.0 ± 10.6	0.063
Male sex	52,155 (69.8%)	7019 (73.9%)	5702 (73.4%)	1409 (76.0%)	45,136 (69.2%)	< 0.001	13,059 (73.7%)	6568 (74.1%)	5368 (73.7%)	1295 (76.0%)	6,491 (73.2%)	0.195
Low income	4,005 (5.4%)	303 (3.2%)	250 (3.2%)	59 (3.2%)	3702 (5.7%)	< 0.001	529 (3.0%)	286 (3.2%)	234 (3.2%)	47 (2.8%)	243 (2.7%)	0.064
CCI (mean ± SD)	3.7 ± 2.1	5.1 ± 2.0	5.0 ± 2.0	5.6 ± 2.1	3.5 ± 2.6	< 0.001	5.1 ± 2.0	5.1 ± 2.0	4.9 ± 2.0	5.5 ± 2.1	5.1 ± 2.0	0.804
CAD	20,975 (28.1%)	6,571 (69.2%)	5263 (67.8%)	1444 (77.8%)	14,404 (22.1%)	< 0.001	12,303 (69.4%)	6097 (68.8%)	4911 (67.5%)	1,319 (77.4%)	6206 (70.0%)	0.078
Stroke	6313 (8.5%)	2610 (27.5%)	1968 (25.4%)	655 (35.3%)	3703 (5.7%)	< 0.001	4550 (25.7%)	2381 (26.9%)	1799 (24.7%)	599 (35.2%)	2169 (21.5%)	< 0.001
DM	14,779 (19.8%)	3604 (38.0%)	2950 (38.0%)	717 (38.7%)	11,175 (17.1%)	< 0.001	6666 (37.6%)	3365 (38.0%)	2775 (38.1%)	654 (38.4%)	3301 (37.2%)	0.329
ESRD	1911 (2.6%)	500 (5.3%)	325 (4.2%)	207 (11.2%)	1411 (2.2%)	< 0.001	885 (5.0%)	451 (5.1%)	293 (4.0%)	185 (10.9%)	434 (4.9%)	0.581
Cancer	7730 (10.3%)	1086 (11.4%)	839 (10.8%)	253 (13.6%)	6,644 (10.2%)	< 0.001	1969 (11.1%)	993 (11.2%)	773 (10.6%)	231 (13.6%)	976 (11.0%)	0.702
AIDS	491 (0.7%)	7 (0.1%)	7 (0.1%)	0 (0%)	484 (0.7%)	< 0.001	10 (0.1%)	7 (0.1%)	7 (0.1%)	0 (0%)	3 (0.03%)	0.343
RA	1004 (1.3%)	127 (1.3%)	101 (1.3%)	33 (1.8%)	877 (1.3%)	0.996	242 (1.4%)	118 (1.3%)	95 (1.3%)	30 (1.8%)	124 (1.4%)	0.746
Psoriasis	705 (0.9%)	114 (1.2%)	78 (1.0%)	36 (1.9%)	591 (0.9%)	0.007	211 (1.2%)	104 (1.2%)	73 (1.0%)	30 (1.8%)	107 (1.2%)	0.890
AS	613 (0.8%)	97 (1.0%)	75 (1.0%)	22 (1.2%)	516 (0.8%)	0.023	173 (1.0%)	92 (1.0%)	70 (1.0%)	22 (1.3%)	81 (0.9%)	0.445
COPD	13,187 (17.6%)	2660 (28.0%)	2135 (27.5%)	568 (30.6%)	10,527 (16.1%)	< 0.001	5045 (28.5%)	2,503 (28.2%)	2,014 (27.7%)	535 (31.4%)	2,542 (28.7%)	0.527
Transplant	135 (0.2%)	24 (0.3%)	17 (0.2%)	8 (0.4%)	111 (0.2%)	0.101	40 (0.2%)	24 (0.3%)	17 (0.2%)	8 (0.5%)	16 (0.2%)	0.268
Pneumoconiosis	69 (0.1%)	13 (0.1%)	11 (0.1%)	**	56 (0.1%)	NS	20 (0.1%)	11 (0.1%)	9 (0.1%)	**	9 (0.1%)	NS
Bronchiectasis	1474 (2.0%)	233 (2.5%)	192 (2.5%)	46 (2.5%)	1241 (1.9%)	< 0.001	462 (2.6%)	216 (2.4%)	177 (2.4%)	42 (2.5%)	246 (2.8%)	0.172
Hypertension	37867 (50.7%)	8688 (91.5%)	7095 (91.4%)	1717 (92.6%)	29179 (44.7%)	< 0.001	16259 (91.7%)	8095 (91.3%)	6646 (91.3%)	1572 (92.2%)	8164 (92.1%)	0.064
TB severity												
Smear positive	30,257 (40.5%)	3445 (36.3%)	2873 (37.0%)	621 (33.5%)	26,812 (41.1%)	< 0.001	6731 (38.0%)	3348 (37.8%)	2795 (38.4%)	604 (35.6%)	3383 (38.2%)	0.599
Culture positive	56,383 (75.4%)	7343 (77.3%)	6030 (77.7%)	1389 (74.9%)	49,040 (75.2%)	< 0.001	13,575 (76.6%)	6784 (76.5%)	5598 (76.9%)	1,263 (74.1%)	6791 (76.6%)	0.915
Cavitation	12,092 (16.2%)	950 (10.0%)	832 (10.7%)	120 (6.5%)	11,142 (17.1%)	< 0.001	1832 (10.3%)	927 (10.5%)	816 (11.2%)	114 (6.7%)	905 (10.2%)	0.604

Abbreviations: AIDS, acquired immunodeficiency syndrome; AS, ankylosing spondylitis; CAD, coronary artery disease; CCI, Charlson comorbidity index; COPD, chronic obstructive pulmonary disease; DM, diabetes mellitus; ESRD, end-stage renal disease; NS, non-significant; PS, propensity score; RA, rheumatoid arthritis; SD, standard deviation; and TB, tuberculosis. * Compared between all antiplatelet users and non-users. ** According to the regulations of National Health Insurance claims database, results with case number less than three are not allowed to be exported.

Table 2. Subgroup analysis of aspirin users (ASA), non-aspirin antiplatelet users (non-ASA), and non-antiplatelet users (non-user) after propensity-score matching.

	ASA (n = 7281)	Matched non-user (n = 7281)	p value	Non-ASA (n = 1704)	Matched non-user (n = 1704)	p value
Age (mean ± SD)	75.4 ± 10.7	75.7 ± 10.7	0.072	77.2 ± 9.9	77.6 ± 9.8	0.244
Male sex	5368 (73.7%)	5317 (73.0%)	0.349	1295 (76.0%)	1275 (74.8%)	0.450
Low income	234 (3.2%)	198 (2.7%)	0.087	47 (2.8%)	43(2.5%)	0.749
CCI	4.9 ± 2.0	5.0 ± 2.0	0.474	5.5 ± 2.1	5.5 ± 2.1	0.634
Coronary artery disease	4911 (67.5%)	4990 (68.5%)	0.166	1319 (77.4%)	1339 (78.6%)	0.432
Stroke	1799 (24.7%)	1647 (22.6%)	0.003	599 (35.2%)	551 (32.3%)	0.089
Diabetes mellitus	2775 (38.1%)	2715 (37.3%)	0.313	654 (38.4%)	649 (38.1%)	0.888
End-stage renal disease	293 (4.0%)	306 (4.2%)	0.617	185 (10.9%)	136 (8.0%)	0.005
Cancer	773 (10.6%)	778 (10.7%)	0.914	231 (13.6%)	210 (12.3%)	0.307
AIDS	7 (0.1%)	3 (0.04%)	0.343	0 (0%)	0 (0%)	> 0.999
Rheumatoid arthritis	95 (1.3%)	96 (1.3%)	> 0.999	30 (1.8%)	27 (1.6%)	0.789
Psoriasis	73 (1.0%)	81 (1.1%)	0.571	30 (1.8%)	26 (1.5%)	0.686
Ankylosing spondylitis	70 (1.0%)	66 (0.9%)	0.796	22 (1.3%)	15 (0.9%)	0.321
COPD	2014 (27.7%)	2051 (28.2%)	0.506	535 (31.4%)	533 (31.3%)	0.971
Transplant	17 (0.2%)	14 (0.2%)	0.719	8 (0.5%)	5 (0.3%)	0.578
Pneumoconiosis	9 (0.1%)	8 (0.1%)	> 0.999	**	**	NS
Bronchiectasis	177 (2.4%)	203 (2.8%)	0.194	42 (2.5%)	47 (2.8%)	0.668
Tuberculosis severity						
Smear positive	2795 (38.4%)	2829 (38.9%)	0.574	604 (35.6%)	609 (35.8%)	0.886
Culture positive	5598 (76.9%)	5602 (76.9%)	0.953	1263 (74.1%)	1272 (74.7%)	0.754
Cavitation	816 (11.2%)	785 (10.8%)	0.427	114 (6.7%)	120 (7.0%)	0.735

Abbreviations: AIDS, acquired immunodeficiency syndrome; COPD, chronic obstructive pulmonary disease; CCI, Charlson comorbidity index; NS, non-significant; and SD, standard deviation. * Compared between all antiplatelet users and non-users. ** According to the regulations of National Health Insurance database, results with case number less than three are not allowed to be exported.

3.3. Survival Analysis

Before PS matching, antiplatelet use was associated with worse overall survival in the univariate analysis (crude hazard ratio (HR): 1.88, 95% confidence interval (CI): 1.82–1.94, $p < 0.0001$) but better overall survival in the multivariate analysis (adjusted HR: 0.90, 95% CI: 0.86–0.93, $p < 0.0001$). After PS matching, antiplatelet use was associated with improved overall survival in the univariate (crude HR: 0.90, 95% CI: 0.87–0.93, $p < 0.0001$) and multivariate analyses (adjusted HR: 0.91, 95% CI: 0.88–0.95, $p < 0.0001$). In the PS-stratified analysis, antiplatelet use was associated with improved overall survival (adjusted HR: 0.91, 95% CI: 0.84–0.97, $p < 0.0001$).

After PS matching, non-ASA antiplatelet use was associated with worse overall survival compared with ASA use (adjusted HR: 1.36, 95% CI: 1.26–1.46, $p < 0.0001$).

As for the comparison of survival between ASA users and antiplatelet non-users, ASA use was associated with improved overall survival (adjusted HR: 0.90, 95% CI: 0.86–0.94, $p < 0.0001$) after PS matching.

As for the comparison of survival between non-ASA antiplatelet users and antiplatelet non-users, there was no association between non-ASA antiplatelet use and overall survival (HR: 1.00, 95% CI: 0.92–1.09, $p = 0.996$) after PS matching.

In the subgroup analysis, while immunocompromised patients had worse survival compared with immunocompetent patients (adjusted HR: 1.47, 95% CI: 1.41–1.54, $p < 0.0001$), antiplatelet use was associated with better survival among both immunocompromised patients (HR: 0.93, 95% CI: 0.89–0.98, $p = 0.010$) and immunocompetent patients (HR: 0.92, 95% CI: 0.86–0.97, $p = 0.003$). Results of the subgroup analysis are illustrated in Figure 2.

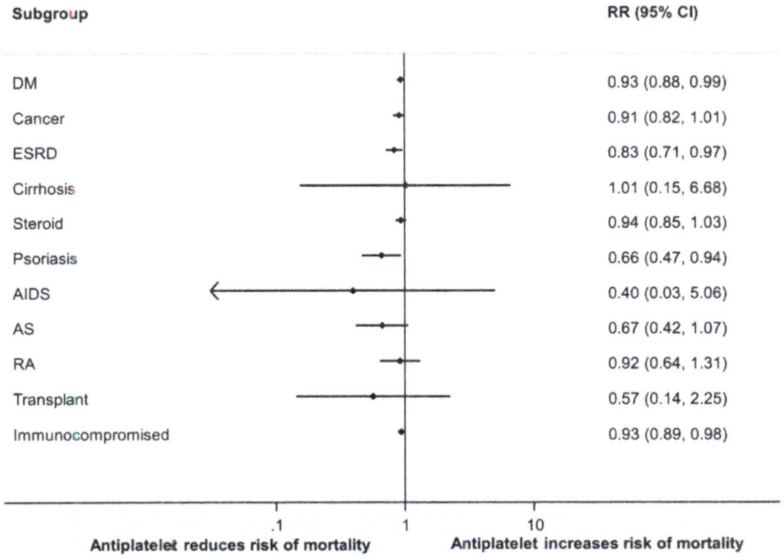

Figure 2. Forest plot of the association between antiplatelet agent use and survival among different groups of patients. Abbreviations: AIDS, acquired immunodeficiency syndrome; AS, ankylosing spondylitis; DM, diabetes mellitus; ESRD, end-stage renal disease; and RA, rheumatoid arthritis

3.4. One-Year Mortality Analysis

Before PS matching, antiplatelet use was associated with a higher one-year mortality rate (crude OR: 1.93, 95% CI: 1.83–2.03, $p < 0.0001$) in the univariate analysis but a lower one-year mortality rate (adjusted OR: 0.91, 95% CI: 0.85–0.96, $p = 0.0015$) in the multivariate analysis. In the PS-stratified analysis, antiplatelet use was also associated with a lower one-year mortality rate (adjusted OR: 0.90, 95% CI: 0.86–0.93, $p < 0.0001$).

After PS matching, antiplatelet use was also associated with a lower one-year mortality rate in the univariate analysis (crude OR: 0.87, 95% CI: 0.82–0.93, $p < 0.0001$) and in the multivariate analysis (adjusted OR: 0.91, 95% CI: 0.85–0.97, $p = 0.004$).

After PS matching, non-ASA antiplatelet use was associated with a higher one-year mortality rate compared with ASA use (adjusted OR: 1.49, 95% CI: 1.32–1.69, $p < 0.0001$).

As for the comparison of survival between ASA users and antiplatelet non-users, ASA use was associated with a lower one-year mortality rate (adjusted OR: 0.90, 95% CI: 0.83–0.97, $p < 0.0001$) after PS matching.

As for the comparison of survival between non-ASA antiplatelet users and antiplatelet non-users, there was no association between non-ASA antiplatelet use and the one-year mortality rate (OR: 0.94, 95% CI: 0.82–1.09, $p = 0.439$) after PS matching.

3.5. Major Bleeding Event Analysis

Incidences of major bleeding were 0.038, 0.035, 0.052, and 0.019 per person-year in antiplatelet users, ASA users, non-ASA antiplatelet users, and antiplatelet non-users, respectively, in the study population before PS matching.

Incidences of major bleeding were 0.037, 0.035, 0.052, and 0.040 per person-year in antiplatelet users, ASA users, non-ASA antiplatelet users, and antiplatelet non-users, respectively, in the study population after PS matching. There was no difference in the bleeding risk between antiplatelet users and non-users in the cumulative incidence function analysis (subhazard ratio: 0.98, 95% CI: 0.90–1.06, $p = 0.604$).

4. Discussion

In our population-based study, we found that antiplatelet use was associated with better overall survival and lower 12-month mortality in pulmonary TB patients who received anti-TB treatment. ASA, which constitutes the majority of antiplatelet agents, provided a survival advantage for TB patients and may be a candidate for auxiliary anti-TB treatment. The safety profile of bleeding events was tolerable among antiplatelet users.

The beneficial effects of antiplatelet agents may result from several mechanisms. First, platelets are known to significantly upregulate monocyte matrix metalloproteinase (MMP)-1 expression, which is associated with lung tissue destruction [10]. In another study, ASA also reduced MMP levels in diabetic rats with induced coronary ischemia [23]. Second, platelets may also act through the modulation of monocyte-derived chemokine and inflammasome activation, leading to a phenotype associated with increased bacterial growth [9]. The increased lung destruction and unrestricted bacterial growth due to platelet activation then leads to poor treatment responses and clinical outcomes. In another murine model study, a low-dose of ASA in combination with other anti-TB agents had anti-inflammatory effects and demonstrated a systemic decrease in neutrophilic recruitment and decreased levels of acute-phase reaction cytokines in the late stage [24]. Interestingly, our study also found that TB disease among antiplatelet users tended not to be cavitated on chest radiography and was also smear negative. This supports antiplatelet agents having beneficial effects of controlling bacterial growth and attenuating lung tissue destruction.

Antiplatelet agents prevent platelet aggregation and are widely used in patients with coronary heart diseases and cerebrovascular events [25]. Different antiplatelet agents inhibit platelet aggregation through different mechanisms. ASA, the most common antiplatelet agent, mainly irreversibly blocks the enzyme cyclooxygenase (COX)-1, thereby preventing the generation of thromboxane A2, a potent platelet activator [26]. Adenosine diphosphate (ADP) antagonists, e.g., clopidogrel, inhibit platelet aggregation by interacting with two purinergic receptors on platelets [27]. A survival advantage was observed in ASA but not in non-ASA antiplatelets. Recently, purinergic pathways were implicated in augmenting the response of killing intracellular pathogens, including *M. tuberculosis*, and also controlling collateral damage [28]. Inhibiting purinergic pathways, as influenced by ADP receptor antagonists, may have adverse impacts on TB treatment. This may explain why a survival benefit was not observed in non-ASA antiplatelet groups. Another explanation may be that non-ASA antiplatelet agents are usually second-line antiplatelets, indicating that non-ASA antiplatelet users may be more disabled or morbid, thus diminishing the benefits of antiplatelet agents. In addition, interestingly, while some basic scientific studies proposed the benefits of ASA, NSAIDs, and even cilostazol in TB treatment, there are no such studies so far that highlighted ADP receptor antagonists as a potential adjunct anti-TB treatment [29–31].

Another important and potential explanation for the finding that overall survival did not improve in non-ASA antiplatelet users when compared to antiplatelet non-users may be that the survival benefit was specific to ASA or NSAID effects but not to anti-platelet effects. ASA can modulate inflammatory processes, such as neutrophil attraction and activation, through small lipid mediators such as thromboxane A2 and prostaglandin E2 through COX pathways [32,33]. Interestingly, NSAIDs have been proposed as an adjunct anti-TB treatment by modulating neutrophil recruitment [34,35]. ASA has also been shown before to reduce neutrophil recruitment in active TB diseases [24]. The abovementioned mechanism may also explain why ASA but not non-ASA antiplatelets were associated with improvement in TB survival.

There have been examples of antiplatelet agents being repurposed for other diseases [26,36,37]. An important instance is the use of antiplatelets in chemoprevention and the treatment of colorectal cancer. It was hypothesized that activated platelets contribute to colorectal tumorigenesis and metastatization via cell–cell interactions and release of tumor mediators [26,36]. A low-dose of ASA therefore prevents tumor growth and metastasis by inhibiting platelet activation during different stages of intestinal tumorigenesis [26,36]. As ASA is an old drug with a well-known safety profile, so repurposing ASA use in anti-TB treatment is an attractive approach to improve TB control.

The TNTR is a web-based national TB notification system maintained by the Taiwanese CDC. Since reporting suspected and confirmed TB diseases to the CDC is mandatory and demanded by law in Taiwan, the completeness and timeliness of the TNTR are excellent. Additional patient and disease characteristics, including smear and culture positivity, are available in the database [38]. Uncertain diagnoses of patients also undergo discussion in an expert meeting to ensure the quality of the diagnoses. Linkage between the TNTR, NHI claims database, and mortality data therefore offered the advantage of high diagnostic accuracy, data abundance, and longitudinal follow-ups [12,39].

Bleeding is a concern with use of antiplatelet agents. A low-dose ASA is defined as 75–325 mg daily, and this is the most widely prescribed dosage of ASA used in cardiovascular diseases, but it still carries an increased risk of bleeding [40]. While the vast majority of ASA tablets in Taiwan contain 75–100 mg of ASA, using a DDD of 90 within 180 days as a cutoff point correlated to a daily dose of around 50–100 mg in our study. The bleeding risk in antiplatelet users was not increased in our study, while the survival benefit remained. A low-dose of ASA may, therefore, be considered if ASA is to be used as an auxiliary therapy for TB.

Other concerns remain regarding the use of antiplatelets in anti-TB regimens in addition to the bleeding risk. Drug–drug interactions may also be another issue. For instance, ASA demonstrated a modest antagonism against isoniazid [41,42]. Rifampicin increases the metabolite of clopidogrel, augments antiplatelet activity, and could be associated with a higher bleeding risk [43]. These safety and therapeutic issues should be taken into consideration and evaluated in future studies.

In our study, the benefit of antiplatelet agents was evident among DM, ESRD, psoriasis, and immunocompromised patients. Indeed, diabetes is associated with increased platelet reactivity, and antiplatelet agents are commonly used for the primary and secondary prevention of cardiovascular events [44,45]. ESRD is also associated with the prothrombotic status of platelet dysfunction [46]. In psoriatic patients, an enhanced cyclooxygenase activity with platelelet hyper-aggregation has been observed [47]. In DM, psoriasis, and ESRD patients who develop active TB, the prothrombotic state may be aggravated. In immunocompromised hosts, delayed and impaired innate immune responses to invasion by *M. tuberculosis* are associated with disease status progression, as in the case of DM [48,49]. Furthermore, the treatment of TB patients with underlying comorbidities is a challenge, and treatment outcomes are usually worse. The beneficial effects of antiplatelets in attenuating thrombosis and immunomodulation in DM, psoriasis, ESRD, and immunocompromised patients may be more evident and worthwhile.

Our study has some limitations. First, we were unable to exclude the possibility of unmeasured confounders. For instance, antiplatelet users may have a higher socioeconomic status (Table 1) and tend to be more health conscious. This may lead to the earlier detection of TB and better treatment outcomes. We, however, included disease severity and a low income in our PS matching. Furthermore, while antiplatelet agents are mainly used for the primary and secondary prevention of cardiovascular and cerebrovascular diseases, which are usually the result of end organ damage of underlying diseases and aging processes, antiplatelet users may be more fragile and disabled than non-users. Second, our study was conducted in Taiwan, the population of which is of Asian ethnicity. Whether these findings can be extrapolated to patients of other ethnicities remains unknown. In addition, this was a retrospective study, and the conclusions drawn from our study can be speculative. Future interventional studies, especially with a randomized design, should be conducted to prove this finding.

In conclusion, our study is the first population-based epidemiological study that has demonstrated a survival benefit among active TB patients receiving antiplatelets. Our study points to a potential direction for designing and discovering new anti-TB agents. Since ASA use was associated with better overall survival and other antiplatelet drugs did not improve survival rates, this phenomenon may suggest that the survival benefit may not be solely due to antiplatelet effects. Additional studies investigating the underlying mechanisms of antiplatelet treatment against TB and co-morbidities would also be of interest.

5. Data Availability

All data are deposited in Department of Statistics, Ministry of Health and Welfare, Taiwan and are not allowed to exported without application and permission.

Supplementary Materials: The following are available online at http://www.mdpi.com/2077-0383/8/7/923/s1, Table S1: Antiplatelet agent category and included drugs, Table S2: Definition of comorbidities and other variables, Table S3: ICD-9-CM and ICD-10-CM codes* used to define intracranial hemorrhage and gastrointestinal bleeding, Table S4: Defined daily doses (DDDs) of drugs in tuberculosis patients with and those without antiplatelet use, Table S5: Defined daily doses (DDDs) of drugs in aspirin users (ASA), non-aspirin antiplatelet users (non-ASA), and non-antiplatelet users (non-users) after propensity-score matching.

Author Contributions: Conceptualization, M.-R.L., C.-H.C., J.-Y.W., and C.-H.L.; data curation, L.-Y.C.; formal analysis, M.-R.L., J.-F.Z., J.-Y.W., and C.-H.L.; funding acquisition, J.-Y.W. and C.-H.L.; investigation, M.-R.L., C.-J.L., and J.-F.Z.; methodology, M.-R.L. and C.-H.L.; project administration, J.-Y.W. and C.-H.L.; software, J.-F.Z.; validation, M.-C.L., C.-H.C., C.-J.L., and L.-Y.C.; writing—original draft, M.-C.L., C.-H.C., L.-Y.C., and C.-H.L.; writing—review & editing, C.-H.C., C.-J.L., and C.-H.L.

Funding: This study was funded by the Taiwan Ministry of Science and Technology (MOST104-2321-B-002-058), the Ministry of Health and Welfare (MOHW105-CDC-C-114-00103; MOHW106-CDC-C-11400104; MOHW107-CDC-C-114-000117), and Wanfang Hospital (108-wf-eva-12). The funders had no role in the study design, data analysis, or manuscript preparation.

Acknowledgments: We thank the Department of Statistics, Ministry of Health and Welfare, Taiwan for maintaining the dataset.

Conflicts of Interest: The authors have no conflict of interest to disclose.

References

1. World Health Organization. *Global Tuberculosis Report*; World Health Organization: Geneva, Switzerland, 2018.
2. World Health Organization. *Resolution WHA67/11: Global Strategy and Targets for Tuberculosis Prevention, Care and Control after 2015*; World Health Organization: Geneva, Switzerland, 2014.
3. Pai, M.; Behr, M.A.; Dowdy, D.; Dheda, K.; Divangahi, M.; Boehme, C.C.; Ginsberg, A.; Swaminathan, S.; Spigelman, M.; Getahun, H.; et al. Tuberculosis. *Nat. Rev. Dis. Primers* **2016**, *2*, 16076. [CrossRef] [PubMed]
4. Belard, S.; Remppis, J.; Bootsma, S.; Janssen, S.; Kombila, D.U.; Beyeme, J.O.; Rossatanga, E.G.; Kokou, C.; Osbak, K.K.; Obiang Mba, R.M.; et al. Tuberculosis Treatment Outcome and Drug Resistance in Lambarene, Gabon: A Prospective Cohort Study. *Am. J. Trop. Med. Hyg.* **2016**, *95*, 472–480. [CrossRef] [PubMed]
5. Lee, C.H.; Wang, J.Y.; Lin, H.C.; Lin, P.Y.; Chang, J.H.; Suk, C.W.; Lee, L.N.; Lan, C.C.; Bai, K.J. Treatment delay and fatal outcomes of pulmonary tuberculosis in advanced age: A retrospective nationwide cohort study. *BMC Infect. Dis.* **2017**, *17*, 449. [CrossRef] [PubMed]
6. Feng, Y.; Dorhoi, A.; Mollenkopf, H.J.; Yin, H.; Dong, Z.; Mao, L.; Zhou, J.; Bi, A.; Weber, S.; Maertzdorf, J.; et al. Platelets direct monocyte differentiation into epithelioid-like multinucleated giant foam cells with suppressive capacity upon mycobacterial stimulation. *J. Infect. Dis.* **2014**, *210*, 1700–1710. [CrossRef] [PubMed]
7. Lee, M.Y.; Kim, Y.J.; Lee, H.J.; Cho, S.Y.; Park, T.S. Mean Platelet Volume in Mycobacterium tuberculosis infection. *BioMed Res. Int.* **2016**, *2016*, 7508763. [CrossRef] [PubMed]
8. Kutiyal, A.S.; Gupta, N.; Garg, S.; Hira, H.S. A Study of Haematological and Haemostasis Parameters and Hypercoagulable State in Tuberculosis Patients in Northern India and the Outcome with Anti-Tubercular Therapy. *J. Clin. Diagn. Res.* **2017**, *11*, OC09–OC13. [CrossRef] [PubMed]
9. Fox, K.A.; Kirwan, D.E.; Whittington, A.M.; Krishnan, N.; Robertson, B.D.; Gilman, R.H.; López, J.W.; Singh, S.; Porter, J.C.; Friedland, J.S. Platelets Regulate Pulmonary Inflammation and Tissue Destruction in Tuberculosis. *Am. J. Respir. Crit. Care Med.* **2018**, *198*, 245–255. [CrossRef]
10. Hortle, E.; Johnson, K.E.; Johansen, M.D.; Nguyen, T.; Shavit, J.A.; Britton, W.J.; Tobin, D.M.; Oehlers, S.H. Thrombocyte Inhibition Restores Protective Immunity to Mycobacterial Infection in Zebrafish. *J. Infect. Dis.* **2019**. [CrossRef]
11. Hsing, A.W.; Ioannidis, J.P. Nationwide Population Science: Lessons from the Taiwan National Health Insurance Research Database. *JAMA Intern. Med.* **2015**, *175*, 1527–1529. [CrossRef]
12. Lin, H.H.; Wu, C.Y.; Wang, C.H.; Fu, H.; Lönnroth, K.; Chang, Y.C.; Huang, Y.T. Association of Obesity, Diabetes, and Risk of Tuberculosis: Two Population-Based Cohorts. *Clin. Infect. Dis.* **2018**, *66*, 699–705. [CrossRef]
13. Lo, H.Y.; Chou, P.; Yang, S.L.; Lee, C.Y.; Kuo, H.S. Trends in tuberculosis in Taiwan, 2002–2008. *J. Formos. Med. Assoc.* **2011**, *110*, 501–510. [CrossRef]
14. Lin, L.Y.; Warren-Gash, C.; Smeeth, L.; Chen, P.C. Data resource profile: The National Health Insurance Research Database (NHIRD). *Epidemiol. Health* **2018**, *40*, e2018062. [CrossRef] [PubMed]
15. Yeh, J.J.; Lin, C.L.; Hsu, C.Y.; Shae, Z.; Kao, C.H. Statin for Tuberculosis and Pneumonia in Patients with Asthma–Chronic Pulmonary Disease Overlap Syndrome: A Time-Dependent Population-Based Cohort Study. *J. Clin. Med.* **2018**, *7*, 381. [CrossRef] [PubMed]
16. Lin, S.Y.; Hsu, W.H.; Lin, C.C.; Lin, C.L.; Tsai, C.H.; Lin, C.H.; Chen, D.C.; Lin, T.C.; Hsu, C.Y.; Kao, C.H. Association of Arrhythmia in Patients with Cervical Spondylosis: A Nationwide Population-Based Cohort Study. *J. Clin. Med.* **2018**, *7*, 236. [CrossRef] [PubMed]
17. Tseng, C.H. Pioglitazone Reduces Dementia Risk in Patients with Type 2 Diabetes Mellitus: A Retrospective Cohort Analysis. *J. Clin. Med.* **2018**, *7*, 306. [CrossRef]
18. Centers for Disease Control, Ministry of Health and Welfare. *Taiwan Guidelines for TB Diagnosis & Treatment*, 6th ed.; Centers for Disease Control, Ministry of Health and Welfare: Taipei, Taiwan, 2017.
19. WHOCC. Definition and General Considerations. 2018. Available online: https://www.whocc.no/ddd/definition_and_general_considera/#Definition (accessed on 26 June 2019).

20. Schulman, S.; Kearon, C. Subcommittee on Control of Anticoagulation of the Scientific and Standardization Committee of the International Society on Thrombosis and Haemostasis. Definition of major bleeding in clinical investigations of antihemostatic medicinal products in non-surgical patients. *J. Thromb. Haemost.* **2005**, *3*, 692–694. [PubMed]
21. Schulman, S.; Angerås, U.; Bergqvist, D.; Eriksson, B.; Lassen, M.R.; Fisher, W. Subcommittee on Control of Anticoagulation of the Scientific and Standardization Committee of the International Society on Thrombosis and Haemostasis. Definition of major bleeding in clinical investigations of antihemostatic medicinal products in surgical patients. *J. Thromb. Haemost.* **2010**, *8*, 202–204.
22. Bannay, A.; Chaignot, C.; Blotiere, P.O.; Basson, M.; Weill, A.; Ricordeau, P.; Alla, F. The Best Use of the Charlson Comorbidity Index with Electronic Health Care Database to Predict Mortality. *Med. Care* **2016**, *54*, 188–194. [CrossRef]
23. Bhatt, L.K.; Veeranjaneyulu, A. Enhancement of matrix metalloproteinase 2 and 9 inhibitory action of minocycline by aspirin: An approach to attenuate outcome of acute myocardial infarction in diabetes. *Arch. Med. Res.* **2014**, *45*, 203–209. [CrossRef]
24. Kroesen, V.M.; Rodriguez-Martinez, P.; Garcia, E.; Rosales, Y.; Díaz, J.; Martín-Céspedes, M.; Tapia, G.; Sarrias, M.R.; Cardona, P.J.; Vilaplana, C. A Beneficial Effect of Low-Dose Aspirin in a Murine Model of Active Tuberculosis. *Front. Immunol.* **2018**, *9*, 798. [CrossRef]
25. Thachil, J. Antiplatelet therapy—A summary for the general physicians. *Clin. Med.* **2016**, *16*, 152–160. [CrossRef] [PubMed]
26. Patrignani, P.; Patrono, C. Aspirin, platelet inhibition and cancer prevention. *Platelets* **2018**, *29*, 779–785. [CrossRef] [PubMed]
27. Angiolillo, D.J. The evolution of antiplatelet therapy in the treatment of acute coronary syndromes: From aspirin to the present day. *Drugs* **2012**, *72*, 2087–2116. [CrossRef] [PubMed]
28. Petit-Jentreau, L.; Tailleux, L.; Coombes, J.L. Purinergic Signaling: A Common Path in the Macrophage Response against Mycobacterium tuberculosis and Toxoplasma gondii. *Front. Cell Infect. Microbiol.* **2017**, *7*, 347. [CrossRef] [PubMed]
29. Maiga, M.; Ammerman, N.C.; Maiga, M.C.; Tounkara, A.; Siddiqui, S.; Polis, M.; Murphy, R.; Bishai, W.R. Adjuvant host-directed therapy with types 3 and 5 but not type 4 phosphodiesterase inhibitors shortens the duration of tuberculosis treatment. *J. Infect. Dis.* **2013**, *208*, 512–519. [CrossRef]
30. Byrne, S.T.; Denkin, S.M.; Zhang, Y. Aspirin and ibuprofen enhance pyrazinamide treatment of murine tuberculosis. *J. Antimicrob. Chemother.* **2007**, *59*, 313–316. [CrossRef]
31. Maitra, A.; Bates, S.; Shaik, M.; Evangelopoulos, D.; Abubakar, I.; McHugh, T.D.; Lipman, M.; Bhakta, S. Repurposing drugs for treatment of tuberculosis: A role for non-steroidal anti-inflammatory drugs. *Br. Med. Bull.* **2016**, *118*, 138–148. [CrossRef]
32. Hinz, C.; Aldrovandi, M.; Uhlson, C.; Marnett, L.J.; Longhurst, H.J.; Warner, T.D.; Alam, S.; Slatter, D.A.; Lauder, S.N.; Allen-Redpath, K.; et al. Human Platelets Utilize Cycloxygenase-1 to Generate Dioxolane A3, a Neutrophil-activating Eicosanoid. *J. Biol. Chem.* **2016**, *291*, 13448–13464. [CrossRef]
33. Schrottmaier, W.C.; Kral, J.B.; Badrnya, S.; Assinger, A. Aspirin and P2Y12 Inhibitors in platelet-mediated activation of neutrophils and monocytes. *Thromb. Haemost.* **2015**, *114*, 478–489. [CrossRef]
34. Vilaplana, C.; Marzo, E.; Tapia, G.; Diaz, J.; Garcia, V.; Cardona, P.J. Ibuprofen therapy resulted in significantly decreased tissue bacillary loads and increased survival in a new murine experimental model of active tuberculosis. *J. Infect. Dis.* **2013**, *208*, 199–202. [CrossRef]
35. Dallenga, T.; Linnemann, L.; Paudyal, B.; Repnik, U.; Griffiths, G.; Schaible, U.E. Targeting neutrophils for host-directed therapy to treat tuberculosis. *Int. J. Med. Microbiol.* **2018**, *308*, 142–147. [CrossRef] [PubMed]
36. Rothwell, P.M.; Price, J.F.; Fowkes, F.G.; Zanchetti, A.; Roncaglioni, M.C.; Tognoni, G.; Lee, R.; Belch, J.F.; Wilson, M.; Mehta, Z.; et al. Short-term effects of daily aspirin on cancer incidence, mortality, and non-vascular death: Analysis of the time course of risks and benefits in 51 randomised controlled trials. *Lancet* **2012**, *379*, 1602–1612. [CrossRef]
37. Ogundeji, A.O.; Pohl, C.H.; Sebolai, O.M. Repurposing of Aspirin and Ibuprofen as Candidate Anti-Cryptococcus Drugs. *Antimicrob. Agents Chemother.* **2016**, *60*, 4799–4808. [CrossRef] [PubMed]
38. Lo, H.Y.; Yang, S.L.; Chou, P.; Chuang, J.H.; Chiang, C.Y. Completeness and timeliness of tuberculosis notification in Taiwan. *BMC Public Health* **2011**, *11*, 915. [CrossRef] [PubMed]

39. Lo, H.Y.; Yang, S.L.; Lin, H.H.; Bai, K.J.; Lee, J.J.; Lee, T.I.; Chiang, C.Y. Does enhanced diabetes management reduce the risk and improve the outcome of tuberculosis? *Int. J. Tuberc. Lung Dis.* **2016**, *20*, 376–382. [CrossRef] [PubMed]
40. Sostres, C.; Lanas, A. Epidemiology of Low Dose Aspirin Damage in the Lower Gastrointestinal Tract. *Curr. Pharm. Des.* **2015**, *21*, 5094–5100. [CrossRef] [PubMed]
41. Byrne, S.T.; Denkin, S.M.; Zhang, Y. Aspirin antagonism in isoniazid treatment of tuberculosis in mice. *Antimicrob. Agents Chemother.* **2007**, *51*, 794–795. [CrossRef] [PubMed]
42. Schaller, A.; Sun, Z.; Yang, Y.; Somoskovi, A.; Zhang, Y. Salicylate reduces susceptibility of Mycobacterium tuberculosis to multiple antituberculosis drugs. *Antimicrob. Agents Chemother.* **2002**, *46*, 2636–2639. [CrossRef]
43. Judge, H.M.; Patil, S.B.; Buckland, R.J.; Jakubowski, J.A.; Storey, R.F. Potentiation of clopidogrel active metabolite formation by rifampicin leads to greater P2Y12 receptor blockade and inhibition of platelet aggregation after clopidogrel. *J. Thromb. Haemost.* **2010**, *8*, 1820–1827. [CrossRef]
44. Park, S.; Kang, H.J.; Jeon, J.H.; Kim, M.J.; Lee, I.K. Recent advances in the pathogenesis of microvascular complications in diabetes. *Arch. Pharm. Res.* **2019**, *42*, 252–262. [CrossRef]
45. Russo, I.; Penna, C.; Musso, T.; Popara, J.; Alloatti, G.; Cavalot, F.; Pagliaro, P. Platelets, diabetes and myocardial ischemia/reperfusion injury. *Cardiovasc. Diabetol.* **2017**, *16*, 71. [CrossRef] [PubMed]
46. Alexopoulos, D.; Panagiotou, A. Oral antiplatelet agents and chronic kidney disease. *Hellenic J. Cardiol.* **2011**, *52*, 509–515. [PubMed]
47. Vila, L.; Cullare, C.; Sola, J.; Puig, L.; de Castellarnau, C.; de Moragas, J.M. Cyclooxygenase activity is increased in platelets from psoriatic patients. *J. Investig. Dermatol.* **1991**, *97*, 922–926. [CrossRef] [PubMed]
48. Lee, M.R.; Huang, Y.P.; Kuo, Y.T.; Luo, C.H.; Shih, Y.J.; Shu, C.C.; Wang, J.Y.; Ko, J.C.; Yu, C.J.; Lin, H.H. Diabetes Mellitus and Latent Tuberculosis Infection: A Systematic Review and Metaanalysis. *Clin. Infect. Dis.* **2017**, *64*, 719–727. [PubMed]
49. Vallerskog, T.; Martens, G.W.; Kornfeld, H. Diabetic mice display a delayed adaptive immune response to Mycobacterium tuberculosis. *J. Immunol.* **2010**, *184*, 6275–6282. [CrossRef] [PubMed]

Journal of Clinical Medicine

MDPI

Article

Factors for the Early Revision of Misdiagnosed Tuberculosis to Lung Cancer: A Multicenter Study in A Tuberculosis-Prevalent Area

Chin-Chung Shu [1,2,†], Shih-Chieh Chang [3,†], Yi-Chun Lai [3], Cheng-Yu Chang [4], Yu-Feng Wei [5,6,†] and Chung-Yu Chen [2,7,*]

1 Department of Internal Medicine, National Taiwan University Hospital, Taipei 100, Taiwan; ccshu@ntu.edu.tw
2 College of Medicine, National Taiwan University, Taipei 100, Taiwan
3 Department of Internal Medicine, National Yang-Ming University Hospital, Yilan County 260, Taiwan; 11319@ymuh.ym.edu.tw (S.-C.C.); toto881049@yahoo.com.tw (Y.-C.L.)
4 Division of Pulmonary Medicine, Department of Internal Medicine, Far Eastern Memorial Hospital, New Taipei City 220, Taiwan; koala2716@hotmail.com
5 Division of Chest Medicine, Department of Internal Medicine, E-Da Hospital/I-Shou University, Kaohsiung 824, Taiwan; yufeng528@gmail.com
6 Institute of Biotechnology and Chemical Engineering, I-Shou University, Kaohsiung 824, Taiwan
7 Division of Pulmonary and Critical Care Medicine, Department of Internal Medicine, National Taiwan University Hospital Yunlin Branch, Yunlin County 640, Taiwan
* Correspondence: c8101147@ms16.hinet.net; Tel.: +886-5-5323911-5675; Fax: +886-5-5335373
† Equal contribution equally to this work.

Received: 2 April 2019; Accepted: 15 May 2019; Published: 17 May 2019

check for updates

Abstract: Background: Lung cancer misdiagnosed as tuberculosis (TB) is not rare, but the factors associated with early diagnosis revision remain unclear. Methods: We screened the cases with TB notification from 2007 to 2018 and reviewed those with misdiagnosis with a revised diagnosis to lung cancer. We analyzed the factors associated with early diagnosis revision (≤1 months) and early obtained pathology (≤1 months) using multivariable Cox regression. Results: During the study period, 45 (0.7%) of 6683 patients were initially notified as having TB, but later diagnosed with lung cancer. The reasons for the original impression of TB were mostly due to image suspicion (51%) and positive sputum acid-fast stain (AFS) (27%). Using multivariable Cox proportional regression, early diagnosis revision was associated with obtaining the pathology early, lack of anti-TB treatment, and negative sputum AFS. Furthermore, the predictors for early obtained pathology included large lesion size (>3 cm), presence of a miliary radiological pattern, no anti-TB treatment, and a culture-negative result when testing for nontuberculous mycobacteria (NTM) using multivariable Cox regression. Conclusion: In patients who are suspected to have TB but no mycobacterial evidence is present, lung cancer should be kept in mind and pathology needs to be obtained early, especially for those with small lesions, radiological findings other than the miliary pattern, and a culture positive for NTM.

Keywords: tuberculosis; lung cancer; misdiagnosis; invasive procedure; revising

1. Introduction

Tuberculosis (TB) remains the most common infectious disease worldwide [1] and, according to the World Health Organization (WHO), an estimated 10.0 million people had active TB, with 1.3 million TB-related deaths reported in 2017 globally [2,3]. Diagnosis optimization is key to providing patient care [4,5]. Although the diagnosis tools regarding TB have been improved in recent decades [6], there are still many cases diagnosed using clinical suspicion without positive culture evidence [7–10] due to time constraints [11]. The most common reasons for a diagnosis based on clinical suspicion are the radiographical or pathological presence of a lesion plus the difficulty and high risk of sample collection other than sputum, or the presence of an acute illness in which TB is suspected without improvement with broad-spectrum antibiotics [7]. In these cases, we might suggest empirical TB treatment and closely follow-up the treatment response. Once a positive treatment response is achieved, the diagnosis of TB can be suggested [7]. However, the accuracy of the tentative diagnosis of TB is still imperfect, with some diagnoses inevitably being incorrect.

Among patients with misdiagnosis, lung cancer is a concern because of its increasing incidence and high mortality [12,13]. In addition, both TB and lung cancer have similar pulmonary manifestations [14,15], such as cavitary lesion, miliary pattern, and pleural effusion [16,17], which can be a trap for clinicians to make a wrong diagnosis. Taiwan has an intermediate prevalence of TB. Differentiation of pulmonary TB from lung cancer can pose a great challenge to clinicians. However, there have been few studies on the timing of invasive diagnostic procedures for diagnosis of pulmonary TB, especially in TB-endemic areas. In a previous study, 1.87% of initial TB notifications are finally diagnosed as lung cancer instead [18]. The disease progression and mortality rate may be increased due to the delay in the cancer diagnosis [19–21]. In addition, TB treatment could induce adverse events in this population with initial misdiagnosis [7,22]. Therefore, in these situations, revising diagnoses early is important, but unclear. We conducted this study to review the patients with an initial notification of TB and then a revised diagnosis to lung cancer in a TB-prevalent area. We aimed to analyze the factors associated with early revision that could improve current patient care.

2. Methods

2.1. Participant Enrollment

This retrospective study was conducted in multiple tertiary referral centers in Taiwan, with one center located on each of the northern, middle, southern, and eastern sides of Taiwan, respectively. Under the approval of the Institutional Review Board of Research Ethics Committee of the study hospitals (NO. 201811047RINA), we screened patients aged ≥20 years that had been diagnosed with tuberculosis by formal notification to Taiwan Centers for Disease Control from January 2007 to August 2018 (details are in the supplementary file). We identified the patients with a final diagnosis of non-TB and among them, newly diagnosised lung cancer responsible for the pulmonary lesion. The patients with pre-existing lung cancer, co-existence of TB and lung cancer, and human immunodeficiency virus infection were excluded (Figure 1).

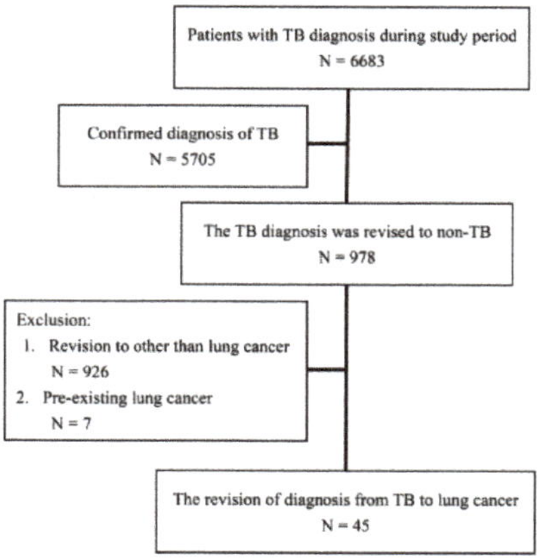

Figure 1. Flow chart of patient recruitment. TB, tuberculosis

2.2. Clinical Information

We retrieved the participants' clinical information, such as age, gender, and details of the initial TB diagnosis from the respective hospital's electronic records. Diagnosis date, anti-TB treatment, and date of diagnosis revision were reviewed in the hospital's record which was the same as that of the Taiwan Centers for Disease Control because TB is a mandatory notification disease in Taiwan. The reasons for initial impression and diagnosis revision were recorded. Radiographic lesions of chest computed tomography (CT) scans were categorized as a nodule/mass, exudative lesion/consolidation, or pleural effusion (Figure 2). In addition, the presence of cavitation and the miliary radiographic pattern and location as well as the size of the main lesion were interpreted by pulmonologists in a default report form. The new diagnosis of lung cancer was defined by a pathology/cytology report and was responsible for the pulmonary lesion suspected to initially be TB. We also reviewed the methods to obtain evidence for cancer diagnosis, the stage of lung cancer, and the mortality.

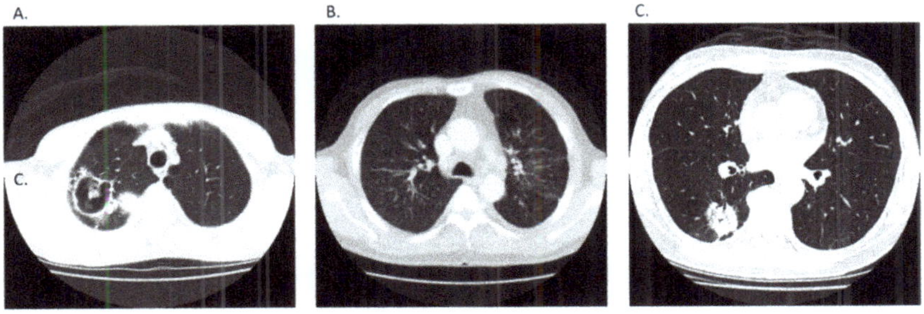

Figure 2. (**A**) Cavitary lung lesion, (**B**) bilateral multiple tiny lung nodules, and (**C**) consolidative lung patch, which all mimicked pulmonary tuberculosis infection [18].

2.3. Outcome and Statistical Analysis

We defined revising the diagnosis early (<1 month) from TB to lung cancer as the primary outcome and obtaining the pathology proof early (<1 month) for lung cancer as the secondary outcome. Inter-group differences were compared using the Mann–Whitney *U* test or one-way ANOVA for continuous variables, where appropriate, and the *Chi* square test for categorical variables. A Kaplan–Meier (KM) curve was used for plotting the time against event curve and was compared using the log rank test. We used Cox proportional hazard regression for analyzing time to diagnosis revision and time to pathology proof obtainment. The clinical, radiological, and microbiological relevant factors were included in the multivariable analysis and the final model was analyzed by the stepwise method. All analyses were performed in SPSS version 19.0 (SPSS Inc, Chicago, IL, USA). All Kaplan–Meier curves were plotted using the Prism software package (GraphPad Version 5.00, San Diego, CA, USA) and analyzed using the log rank test. A 2-tailed *p* value of <0.05 was considered significant.

3. Results

3.1. Participant Demographics and Reasons for TB Suspicions

During the study period, a total of 6683 patients had been diagnosed in the study hospitals. Among them, the diagnoses of 978 (14.6%) and 45 (0.7%) had been revised to non-TB causes and lung cancer, respectively (Figure 1). The average age was 66 years and 71% were male (Table 1). Acid-fast stain (AFS) was positive for 20% and the lesion size was around 4.7 cm. Pleural effusion and radiological findings of cavitary and miliary patterns accounted 20%, 22%, and 11%, respectively. The reasons for tentatively diagnosed TB included suspicion of radiological findings (*n* = 23, 51%), positive AFS (*n* = 12, 27%), pathology suspicion for TB (*n* = 2, 4%), lymphocyte-predominant pleural effusion (n = 2, 4%), and other clinical suspicion (n = 8, 18%). Although all these patients were diagnosed with TB, anti-TB treatment was only applied only in 82%. The average length of time to obtain pathology and revise the diagnosis were 52.9 ± 74.4 days and 57.6 ± 44.7 days, respectively.

Table 1. The demographics of patients according to the time to revise the diagnosis.

	All (n = 45)	Early Revision (n = 17)	Late Revision (n = 28)	p Value
Age (years)	66.1 ± 13.3	69.3 ± 11.8	64.1 ± 13.9	0.242
Male sex	32 (71%)	11 (75%)	21 (75%)	0.460
Microbiology				
AFS *				0.368
Positive	9 (20%)	4 (24%)	5 (18%)	
Negative	35 (78%)	12 (71%)	23 (82%)	
Culture *				0.147
NTM	8 (18%)	4 (24%)	4 (14%)	
Negative	16 (36%)	3 (18%)	13 (46%)	
Chest CT radiographic pattern				0.529
Nodular	28 (62%)	11 (65%)	17 (61%)	
Consolidation	15 (33%)	6 (35%)	9 (32%)	
Pleural effusion	2 (4%)	0	2 (7%)	
Lesion size (cm)	4.7 ± 2.8	5.2 ± 3.0	4.4 ± 2.7	0.235
Bilateral	6 (13%)	3 (18%)	3 (11%)	0.710
Cavitation	10 (22%)	1 (6%)	9 (32%)	0.040
Miliary pattern	5 (11%)	3 (18%)	2 (7%)	0.277
Pleural effusion amount				0.388
<1/3 hemithorax	3 (7%)	2 (12%)	1 (4%)	
1/3–2/3 hemithorax	5 (11%)	2 (12%)	3 (11%)	
>2/3 hemithorax	1 (2%)	1 (6%)	0	
The cause for TB suspicion				0.732
AFS (+)	12 (27%)	5 (29%)	7 (25%)	
Pathology suspicion	2 (4%)	0	2 (7%)	
Radiological suspicion	23 (51%)	9 (53%)	14 (50%)	
Clinical suspicion	8 (18%)	3 (18%)	5 (18%)	
Anti-TB treatment	37 (82%)	11 (65%)	26 (93%)	0.017
Days to obtain pathology	52.9 ± 74.4	46.9 ± 104.0	56.5 ± 50.7	0.009
Days to diagnosis revision	57.6 ± 44.7	15.1 ± 9.4	83.4 ± 37.0	<0.001

Abbreviations: AFS, acid-fast smear; CT, Computed Tomography; TB, tuberculosis; NTM, nontuberculous mycobacteria. * One case and 21 cases were not tested for AFS and mycobacteria culture, respectively.

3.2. Final Diagnosis of Lung Cancer Status

The pathology proof was obtained using CT-guidance in 14 patients (31%), bronchoscopy in 11 patients (24%), echo-guidance in 9 patients (20%), surgical procedure in 9 patients (20%), and pleural/pericardial effusion in 2 patients (4%). In regard to the cancer stage, there were 26 patients (58%) with metastases (stage 4) in the final diagnosis of lung cancer. In addition, 8 patients (18%) were stage 1, 3 patients (7%) were stage 2, and 8 patients (18%) were stage 3. Regarding the cancer type, 59% of patients had adenocarcinoma, 12% had squamous cell carcinoma, 9% had small cell carcinoma, 15% had poorly-differentiated carcinoma, 1 had pleural cancer, and 1 had sarcomatoid carcinoma. Nineteen patients (42%) died within 1 year follow-up.

3.3. The Factors Favored Early Revising Diagnosis

For patients with early diagnosis revision (n = 17, 32%) (Table 1), their age and demographics were similar with those with late revision except they had less cavitation pattern (6% vs. 32%, p = 0.040) and anti-TB treatment (65% vs 93%, p = 0.017) as well as lower time to obtain pathology (46.9 days vs. 56.5 days, p = 0.009). In multivariable Cox proportion analysis, early obtaining pathology was an independent factor associated with early diagnosis revision (Hazard ratio (HR):

2.079 (1.041–4.154), $p = 0.038$). Anti-TB treatment (HR: 0.363 (0.120–1.097), $p = 0.073$) and positive AFS (HR: 0.409 (0.146–1.148), 0.089) were borderline associated with diagnosis revision negatively (Table 2). The Kaplan-Meier (KM) curve also showed that early obtaining pathology could be significantly correlated with diagnosis revision earlier (Figure 3, $p = 0.015$).

Table 2. Multi-variable analysis for early diagnosis revision.

Characteristics	Multivariate	
	HR (95% C.I.)	*p* Value
Early obtaining pathology vs late group	2.079 (1.041–4.154)	0.038
Empirical anti-TB treatment vs none	0.363 (0.120–1.097)	0.073
AFS positive vs negative/not done	0.409 (0.146–1.148)	0.089

Abbreviation: AFS, acid-fast smear; TB, tuberculosis. All clinical, radiological, and microbiological factors were used for multivariable Cox proportional regression using stepwise methods.

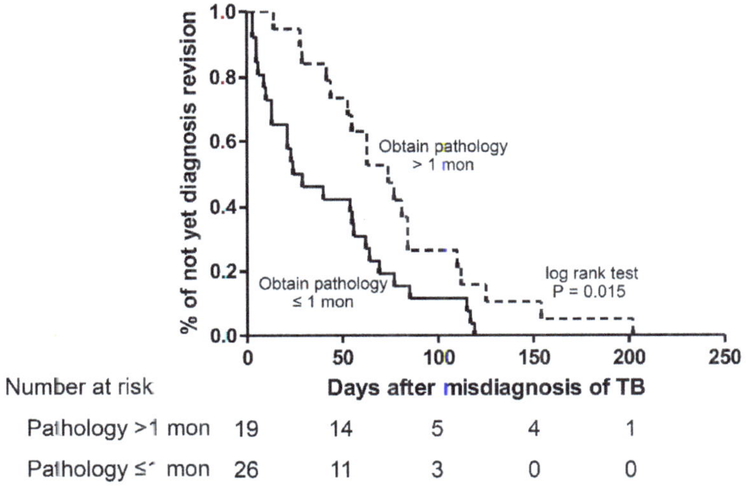

Figure 3. Kaplan Meier curve for time to revise diagnosis to lung cancer according to time to obtain pathology proof.

3.4. The Factors Associated with Early Obtained Pathology

We aimed to study the factors correlated with obtaining pathology proof early ($n = 26$, 58%). The patients with early obtained pathology had similar age, sex, positive proportion of sputum AFS, radiological pattern, and diagnosis evidence in comparison to those with late obtained pathology (Table S1 in supplementary file). In contrast, larger mass size (5.8 vs. 2.7 cm, $p < 0.001$), more miliary pattern (19% vs. 0%, 0.043), less anti-TB treatment (69% vs. 100%, $p = 0.008$), and shorter time to diagnosis revision (42.2 vs 78.6 days, $p = 0.006$) were found in the early pathology group than in late group.

We validated associated factors for early pathology by multivariable Cox proportional regression analysis (Table 3). The analysis showed that lesion size (HR: 4.258 (1.659–10.924) per 1 cm increment, $p = 0.003$), miliary radiographic pattern (HR:14.739 (4.096–53.038), $p = 0.001$), empirical anti-TB treatment (HR: 0.305 (0.094–0.994), $p = 0.049$), and sputum culture positive for nontuberculous mycobacteria (NTM) (HR: 0.310 (0.114–0.841), $p = 0.021$) were independent factors. The four independent factors were shown to be statistically significant by KM curves (Figure 4).

Table 3. Multi-variable analysis for early obtained pathology.

Characteristics	Multivariate	
	HR (95% C.I.)	*p* Value
Miliary radiographic pattern vs. others	14.739 (4.096–53.038)	<0.001
Lesion size per 1 cm increment	4.258 (1.659–10.924)	0.003
Empirical anti-TB treatment vs. no treatment	0.305 (0.094–0.994)	0.049
Culture (+) for NTM vs. negative and not done	0.310 (0.114–0.841)	0.021

Abbreviations: TB, tuberculosis; NTM, nontuberculous mycobacteria. All clinical, radiological, and microbiological factors were used for multivariable Cox proportional regression using stepwise methods.

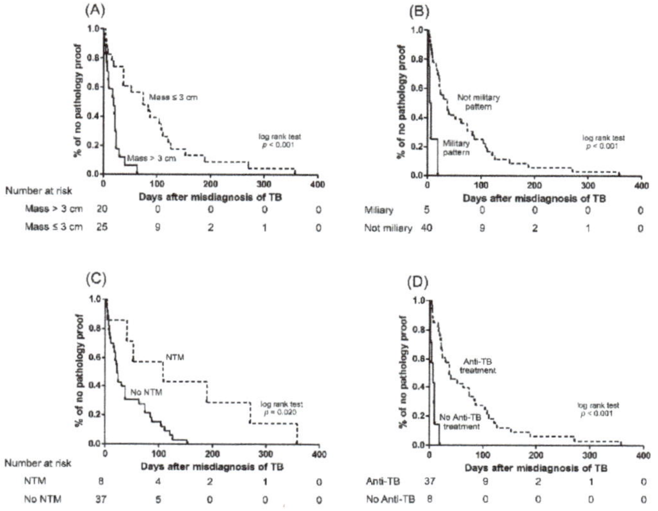

Figure 4. Kaplan–Meier curves for time to obtain pathology proof of lung cancer according to (**A**) lesion size, (**B**) radiographic miliary pattern, (**C**) mycobacterial culture, and (**D**) anti-tuberculosis treatment. NTM, nontuberculous mycobacteria; TB, tuberculosis.

4. Discussion

In the present study, the incidence of patients with misdiagnosis of TB was around 14.6% with some revised early to obtain a diagnosis of lung cancer (0.7%). We conducted this study to review the patients with initial notification of TB and then diagnosis revision to lung cancer in a TB-prevalent area. This study found that obtaining pathology early could help diagnoses be revised earlier. In particular, patients with a lesion size >3 cm, a radiologic miliary pattern, no anti-TB treatment, and cultures negative for NTM could alert clinicians to arrange a pathology exam. In contrast, those with a lesion size ≤3 cm, no radiologic miliary pattern, anti-TB treatment, and cultures positive for NTM are prone to have a delayed diagnosis and our conclusion emphasizes that lung cancer should be kept in mind in this subgroup and earlier pathology should be obtained if feasible.

The clinical and radiological manifestations of lung cancer and tuberculosis are similar in some aspects [14,15]. For example, they are chronic and could have mass lesions, cavitation, and multiple pulmonary lesions [16,17]. Many lung abnormalities, such as nodules or masses discovered by chest radiography or CT scan, are considered suspicious enough to prompt immediate biopsy. In certain clinical situations, it may be desirable to further characterize a pulmonary nodule by imaging. Some investigations found that the CT characteristics of lung nodules or masses helped in the differentiation of benign and malignant pulmonary lesions [19]. However, some patients are at high risk of complications during invasive procedures, such as biopsy and bronchoscopy, so clinicians might try anti-TB treatment and follow the response to judge the diagnosis [7]. Therefore, TB could be misdiagnosed in a patient actually having lung cancer. However, this could lead to a delayed lung cancer diagnosis and late cancer treatment, which could lead to poorer outcomes [20–22]. In addition, anti-TB treatment would be unnecessary and its adverse effect and cost could be harmful for patients [7,23]. In such a misdiagnosis, how to revise a diagnosis early is very important, especially in a TB-prevalent area [24].

In the present study, early obtained tissue proof is a direct factor associated with shorter time to diagnosis revision to lung cancer; it is easy to understand this causal relationship. Similar findings have been found in our previous observation [18] but no study repeated the findings until the present investigation. Although the misdiagnosis rate decreased, the harm is still present and needs to be improved because it matters in delaying cancer diagnosis and influences prognosis. For patients with positive sputum AFS, the diagnosis revision might be postponed. Because concomitant cancer and infection is not uncommon [25,26], we would not totally exclude mycobacterial infection before we yielded negative cultures for *Mycobacterium tuberculosis*, which is time consuming [11]. In regard to patients with anti-TB treatment, the diagnosis revision is also late. This might be explained by the fact that the overall status is highly suspected to TB, so the diagnosis revision would not be made until the patient did not respond to the course of treatment. However, the treatment response for TB is usually not fast [27,28].

Therefore, to obtaining pathology early would be a key step to decrease the time to revise the diagnosis to lung cancer. There were four factors that significantly predicted early obtained pathology. First, the size of the radiological lesion is important because large lesion always alert clinician more and the treatment response would be easily to follow. Once it is clear the patient is unresponsive to treatment and the lesion is huge, i.e., >3 cm, the clinician will favor further tissue proof if no contraindications. Second, the miliary pattern can exist in both TB and lung cancer [29,30]. However, the miliary pattern in lung cancer usually shows bigger, multiple nodules and indicates terminal cancer with metastasis, which hints to physicians to think of cancer instead of TB. Third, no anti-TB regimen and cultures negative for NTM indicate a low chance of mycobacterial infection, so the tie to obtain pathology might be early in clinical practice. In fact, even if sputum culture is positive for NTM, the clinical colonization of NTM is not low [31], so the diagnosis needs to be re-considered seriously. In contrary, those with late obtained pathology require us to pay attention; they usually have small lesions, receive anti-TB treatment, have cultures positive for NTM, and no miliary pattern. However, the validation for the four-factor model is needed in the future.

Several limitations existed in this study. First, this is a retrospective study and different groups are not identical in demographics. In addition, no standard protocol was implemented for the procedure to obtain pathology. Among the technical part, radiology follow-up, CT acquisition protocol, and post-processing are missing. Slice acquisition, respiratory artifacts, and dose-length product (DLP) radiation dose all could have affected the results. All these elements would be required to determine the potential clinical impact of this study. Third, the case number is small, although we did review the cases from four medical referral centers. If the parent population focused on TB notification cases, it would underestimate the percentage of lung cancer mimicking TB.

In conclusion, we found there is small proportion (0.7%), but not a rare amount, of cases of lung cancer initially misdiagnosed as TB. Under such misdiagnosis, obtaining pathology early is a solution to revising the diagnosis early. For those with small lesions, radiological findings other than miliary pattern, receiving empirical anti-TB treatment, and sputum cultures positive for NTM, the time to obtain pathology is longer and we should carefully monitor the treatment response and dynamic lesion change in these patients. Further tissue proof is suggested, especially in patients without obvious treatment response or definite TB evidence.

Ethics approval and consent to participate: The Research Ethics Committee of National Taiwan University Hospital approved this study (IRB No.: 201811047RINA).

Supplementary Materials: The following are available online at http://www.mdpi.com/2077-0383/8/5/700/s1, Table S1: The demographics of patients according to the timing to obtain pathology proof.

Author Contributions: C.-C.S., S.-C.C., Y.-C.L., C.-Y.C. (Cheng-Yu Chang), Y.-F.W. and C.-Y.C. (Chung-Yu Chen) involved data collection, interpretation, analysis and manuscript writing. C.-Y.C. (Chung-Yu Chen) is responsible for the coordination.

Acknowledgments: The authors thank staff of infection control center and TB manager in National Taiwan University Hospital and its Yunlin Branch, National Yang-Ming University Hospital, Far Eastern Memorial Hospital and E-Da Hospital for their support with data and statistics.

Conflicts of Interest: All authors declare no financial, professional or other personal interest of any nature or kind in a related product, service, and/or company.

References

1. World Health Organization. *Group at Risk: WHO Report on the Tuberculosis Epidemic*; World Health Organization: Geneva, Switzerland, 1996.
2. *Global Tuberculosis Report 2018*; World Health Organization: Geneva, Switzerland, 2018.
3. MacNeil, A.; Glaziou, P.; Sismanidis, C.; Maloney, S.; Floyd, K. Global epidemiology of tuberculosis and progress toward achieving global targets—2017. *MMWR Morb. Mortal. Wkly Rep.* **2019**, *68*, 263–266. [CrossRef] [PubMed]
4. WHO. The End TB Strategy. 2015. Available online: https://www.who.int/tb/post2015_strategy/en/.
5. Lönnroth, K.; Migliori, G.B.; Abubakar, I.; D'Ambrosio, L.; de Vries, G.; Diel, R.; Douglas, P.; Falzon, D.; Gaudreau, M.A.; Goletti, D.; et al. Towards tuberculosis elimination: an action framework for low-incidence countries. *Eur. Respir. J.* **2015**, *45*, 928–952. [CrossRef]
6. Goletti, D.; Lee, M.R.; Wang, J.Y.; Walter, N.; Ottenhoff, T.H.M. Update on tuberculosis biomarkers: From correlates of risk, to correlates of active disease and of cure from disease. *Respirology* **2018**, *23*, 455–466. [CrossRef]
7. Chiang, C.Y.W.J.; Yu, M.C.; Lee, J.J.; Lee, P.I.; Lee, P.H.; Chou, J.H.; Lin, H.H.; Chiang, I.H.; Hung, C.C.; So, R.; et al. *Taiwan Guidelines for TB Diangosis and Treatment*, 5th ed.; Center for Disease Control, Executive Yuan: Taipei, Taiwan, 2017.
8. Rozenshtein, A.; Hao, F.; Starc, M.T.; Pearson, G.D. Radiographic appearance of pulmonary tuberculosis: Dogma disproved. *AJR Am. J. Roentgenol.* **2015**, *204*, 974–978. [CrossRef]
9. Curley, C.A. Rule out pulmonary tuberculosis: Clinical and radiographic clues for the internist. *Cleve. Clin. J. Med.* **2015**, *82*, 32–38. [CrossRef] [PubMed]
10. Masamba, L.P.L.; Jere, Y.; Brown, E.R.S.; Gorman, D.R. Tuberculosis diagnosis delaying treatment of cancer: Experience from a New Oncology Unit in Blantyre, Malawi. *J. Glob. Oncol.* **2016**, *2*, 26–29. [CrossRef]
11. Lu, D.; Heeren, B.; Dunne, W.M. Comparison of the Automated Mycobacteria Growth Indicator Tube System (BACTEC 960/MGIT) with Lowenstein-Jensen medium for recovery of mycobacteria from clinical specimens. *Am. J. Clin. Pathol.* **2002**, *118*, 542–545. [CrossRef] [PubMed]
12. Chiang, C.J.; Lo, W.C.; Yang, Y.W.; You, S.L.; Chen, C.J.; Lai, M.S. Incidence and survival of adult cancer patients in Taiwan, 2002–2012. *J. Formos. Med. Assoc.* **2016**, *115*, 1076–1088. [CrossRef]
13. Jung, K.W.; Won, Y.J.; Oh, C.M.; Kong, H.J.; Lee, D.H.; Lee, K.H. Cancer statistics in Korea: Incidence, mortality, survival, and prevalence in 2014. *Cancer Res. Treat.* **2017**, *49*, 292–305. [CrossRef] [PubMed]
14. Liu, Y.; Wang, H.; Li, Q.; McGettigan, M.J.; Balagurunathan, Y.; Garcia, A.L.; Thompson, Z.J.; Heine, J.J.; Ye, Z.; Gillies, R.J.; et al. Radiologic features of small pulmonary nodules and lung cancer risk in the National Lung Screening Trial: A nested case-control study. *Radiology* **2018**, *286*, 298–306. [CrossRef]
15. Nachiappan, A.C.; Rahbar, K.; Shi, X.; Guy, E.S.; Barbosa, E.J.M.; Shroff, G.S.; Ocazionez, D.; Schlesinger, A.E.; Katz, S.I.; Hammer, M.M. Pulmonary tuberculosis: Role of radiology in diagnosis and management. *Radiographics* **2017**, *37*, 52–72. [CrossRef]
16. Light, R.W. Update on tuberculous pleural effusion. *Respirology* **2010**, *15*, 451–458. [CrossRef]
17. Light, R.W. Clinical practice. Pleural effusion. *N. Engl. J. Med.* **2002**, *346*, 1971–1977. [CrossRef]
18. Chen, C.Y.; Wang, J.Y.; Chien, Y.C.; Chen, K.Y.; Yu, C.J.; Yang, P.C. Lung cancer mimicking pulmonary tuberculosis in a TB-endemic country: The role of early invasive diagnostic procedures. *Lung Cancer Manag.* **2015**, *4*, 9–16. [CrossRef]

19. Lobrano, M.B. Partnerships in oncology and radiology: The role of radiology in the detection, staging, and follow-up of lung cancer. *Oncologist* **2006**, *11*, 774–779. [CrossRef]
20. Burki, T.K. Late detection of lung cancer. *Lancet Oncol.* **2014**, *15*, e590. [CrossRef]
21. Vinas, F.; Ben Hassen, I.; Jabot, L.; Monnet, I.; Chouaid, C. Delays for diagnosis and treatment of lung cancers: A systematic review. *Clin. Respir. J.* **2016**, *10*, 267–271. [CrossRef] [PubMed]
22. Mohammed, N.; Kestin, L.L.; Grills, I.S.; Battu, M.; Fitch, D.W.; Wong, C.O.; Margolis, J.H.; Chmielewski, G.W.; Welsh, R.J. Rapid disease progression with delay in treatment of non-small-cell lung cancer. *Int. J. Radiat. Oncol. Biol. Phys.* **2011**, *79*, 466–472. [CrossRef] [PubMed]
23. Shu, C.C.; Lee, C.H.; Lee, M.C.; Wang, J.Y.; Yu, C.J.; Lee, L.N. Hepatotoxicity due to first-line anti-tuberculosis drugs: A five-year experience in a Taiwan medical centre. *Int. J. Tuberc. Lung Dis.* **2013**, *17*, 934–939. [CrossRef]
24. Centers of Disease Control DoH, R.O.C. (Taiwan). *CDC Annual Report 2018*; Centers of Disease Control, Department of Health: Taipei, Taiwan, 2018.
25. Morales-Garcia, C.; Parra-Ruiz, J.; Sanchez-Martinez, J.A.; Delgado-Martin, A.E.; Amzouz-Amzouz, A.; Hernandez-Quero, J. Concomitant tuberculosis and lung cancer diagnosed by bronchoscopy. *Int. J. Tuberc. Lung Dis.* **2015**, *19*, 1027–1032. [CrossRef]
26. Varol, Y.; Varol, U.; Unlu, M.; Kayaalp, I.; Ayranci, A.; Dereli, M.S.; Guclu, S.Z. Primary lung cancer coexisting with active pulmonary tuberculosis. *Int. J. Tuberc. Lung Dis.* **2014**, *18*, 1121–1125. [CrossRef] [PubMed]
27. Seon, H.J.; Kim, Y.I.; Lim, S.C.; Kim, Y.H.; Kwon, Y.S. Clinical significance of residual lesions in chest computed tomography after anti-tuberculosis treatment. *Int. J. Tuberc. Lung Dis.* **2014**, *18*, 341–346. [CrossRef] [PubMed]
28. Su, W.J.; Feng, J.Y.; Chiu, Y.C.; Huang, S.F.; Lee, Y.C. Role of 2-month sputum smears in predicting culture conversion in pulmonary tuberculosis. *Eur. Respir. J.* **2011**, *37*, 376–383. [CrossRef] [PubMed]
29. Salahuddin, M.; Karanth, S.; Ocazionez, D.; Estrada, Y.M.R.M.; Cherian, S.V. Clinical characteristics and etiologies of miliary nodules in the US: A single-center study. *Am. J. Med.* **2019**. [CrossRef] [PubMed]
30. Furqan, M.; Butler, J. Miliary pattern on chest radiography: TB or not TB? *Mayo Clin. Proc.* **2010**, *85*, 108. [CrossRef] [PubMed]
31. Van Ingen, J.; Bendien, S.A.; de Lange, W.C.M.; Hoefsloot, W.; Dekhuijzen, P.N.R.; Boeree, M.J.; van Soolingen, D. Clinical relevance of non-tuberculous mycobacteria isolated in the Nijmegen-Arnhem region, The Netherlands. *Thorax* **2009**, *64*, 502–506. [CrossRef] [PubMed]

Journal of
Clinical Medicine

MDPI

Review

Type 2 Diabetes Mellitus and Altered Immune System Leading to Susceptibility to Pathogens, Especially *Mycobacterium tuberculosis*

Steve Ferlita [1], Aram Yegiazaryan [2], Navid Noori [1], Gagandeep Lal [1], Timothy Nguyen [1], Kimberly To [3] and Vishwanath Venketaraman [1,2,3,*]

1 College of Osteopathic Medicine of the Pacific, Western University of Health Sciences, Pomona, CA 91766-1854, USA; steve.ferlita@westernu.edu (S.F.); nnoori@westernu.edu (N.N.); gagandeep.lal@westernu.edu (G.L.); timothy.nguyen@westernu.edu (T.N.)

2 Graduate College of Biomedical Sciences, Western University of Health Sciences, Pomona, CA 91766-1854, USA; aram.yegiazaryan@westernu.edu

3 Department of Basic Medical Sciences, College of Osteopathic Medicine of the Pacific, Western University of Health Sciences, Pomona, CA 91766-1854, USA; kimberly.to@westernu.edu

* Correspondence: vvenketaraman@westernu.edu; Tel.: +1-909-706-3736; Fax: +1-909-469-5698

Received: 18 November 2019; Accepted: 10 December 2019; Published: 16 December 2019

check for updates

Abstract: There has been an alarming increase in the incidence of Type 2 Diabetes Mellitus (T2DM) worldwide. Uncontrolled T2DM can lead to alterations in the immune system, increasing the risk of susceptibility to infections such as *Mycobacterium tuberculosis (M. tb)*. Altered immune responses could be attributed to factors such as the elevated glucose concentration, leading to the production of Advanced Glycation End products (AGE) and the constant inflammation, associated with T2DM. This production of AGE leads to the generation of reactive oxygen species (ROS), the use of the reduced form of nicotinamide adenine dinucleotide phosphate (NADPH) via the Polyol pathway, and overall diminished levels of glutathione (GSH) and GSH-producing enzymes in T2DM patients, which alters the cytokine profile and changes the immune responses within these patients. Thus, an understanding of the intricate pathways responsible for the pathogenesis and complications in T2DM, and the development of strategies to enhance the immune system, are both urgently needed to prevent co-infections and co-morbidities in individuals with T2DM.

Keywords: type 2 diabetes; co-morbidities; co-infections; cytokines; inflammation; redox imbalance; antioxidants

1. Introduction

1.1. *Type 2 Diabetes Mellitus*

Type 2 Diabetes Mellitus is a prevalent disease throughout the world. The World Health Organization (WHO) reports that an estimated 422 million people all over the world are living with diabetes. The WHO projects that diabetes will be the seventh leading cause of death in 2030 [1]. In the United States, the Center for Disease Control (CDC) reports that 30.3 million people have diabetes, with approximately 23.1 million people diagnosed every year and a staggering 84.1 million Americans in pre-diabetic stage [2]. Tuberculosis (TB) is a serious threat for people living with diabetes. In 2017, 20% of people with TB in the United States also had diabetes. Currently, one third of the population of the world is infected with latent *Mycobacterium tuberculosis (M. tb)* [3]. At this stage, the individual is not infectious; however, TB can be reactivated as a result of granuloma liquefaction due to immunodeficiency or immunocompromisation [4], as in the case with individuals suffering from

Type 2 Diabetes Mellitus (T2DM) [5]. Due to the high prevalence of diabetes worldwide, necessary measures must be taken to understand, prevent, and treat T2DM.

Insulin, a peptide hormone produced by beta cells within the pancreas, facilitates the absorption of glucose into cells from blood, in order to maintain proper glucose levels [6]. T2DM is a disease characterized by the inability of the body to produce sufficient insulin, or the development of insulin resistance [7]. As a result, T2DM's state of insufficient insulin production or insulin-resistance can have detrimental complications, including macrovascular diseases such as hypertension, coronary artery disease, heart attacks and strokes; or microvascular diseases such as neuropathy, nephropathy, and cancer [8]. In order to effectively reduce the incidence of T2DM worldwide and prevent concurrent infections, efforts must be undertaken to discern the etiology, to create rationale treatments for symptomatic patients, and to develop preventive measures.

1.2. Pathogenesis in T2DM

Diabetes complications resulting from an impaired glucose metabolism have been associated with retinopathy, nephropathy, and polyneuropathy [9]. When present, excess glucose is not oxidized and is shunted to the polyol pathway, consisting of two enzymes—aldose reductase (AR) and sorbitol dehydrogenase (SDH) [10]. AR reduces glucose to sorbitol in the presence of its co-factor, NADPH; SDH, with its co-factor nicotinamide adenine dinucleotide (NAD+), converts sorbitol to fructose—a more potent nonenzymatic glycation agent than glucose. Thus, an increase in glucose flux through the polyol pathway (as happens in T2DM) will decrease the available NADPH and increase the production of Advanced Glycation End products (AGE), leading to oxidative stress in a pathway we discuss later in this review [11]. The SDH oxidation of sorbitol to fructose also causes oxidative stress due to the co-factor conversion of NAD+ to NADH, the substrate for NADH oxidase which generates reactive oxygen species (ROS) [12]. Elevated sorbitol levels have also been associated with cellular and organ damage by directly depleting myoinositol (MI). This MI deficiency alters the metabolism of phosphoinositides and reduces diacylglycerol (DAG), inositol triphosphate (IP3) and protein kinase C (PKC) activation, resulting in a reduction in NA^{+}/K^{+} ATPase pump activity. Metabolic pathway derangement causes conductance abnormalities, resulting in neurological complications such as the neuropathy observed in diabetic patients [10]. Additionally, SDH enzyme deficiency in nervous tissues, the kidney, and the lens and retina of the eye lead to the elevation of sorbitol, leading to the development and progression of retinopathy, cataracts and neuropathy observed in diabetic patients [13].

Concomitantly, the cellular antioxidant capacity of Glutathione (GSH)is diminished due to AR activity, which depletes the NADPH required for glutathione reductase (GSR) to recycle glutathione (GSH). A reduction in available NADPH, due to the overutilization of the Polyol Pathway, significantly decreases GSH levels and thus limits the ROS-scavenging activity of GSH, and further increases ROS levels [14]. A diabetic mouse model (MKR) vs. control study performed in 2010 revealed sorbitol levels 2.5 times higher in MKR when compared to a control Also found was 1.7 times lower levels of the reduced glutathione (rGSH) in skeletal muscle [15]. Therefore, it is understood that the overutilization of the polyol pathway due to glucose excess is a major source of oxidative stress, by limiting the recycling of GSH via GSR induced by diabetes.

1.3. The Production of Glutathione in T2DM

GSH synthesis occurs intracellularly in a two-step enzymatic reaction. First, glutamine is joined with cysteine via the rate-limiting enzyme glutamine–cysteine ligase (GCLC) making γ–glutamyl cysteine. The second enzyme required for GSH synthesis is glutathione synthetase (GSS), which links γ–glutamyl cysteine to glycine to form the functional molecule of GSH. Once formed, GSH can perform its function by absorbing and detoxifying reactive oxygen species to prevent cellular damage. GSH can be found in vivo in two forms—reduced, functional glutathione (rGSH/GSH), and oxidized glutathione (GSSG). GSSG can be recycled to form rGSH/GSH via the enzyme GSR which, as previously stated, requires NADPH [16].

GSH levels have been shown to be significantly compromised in individuals with T2DM, due to a diminishment in the levels of NADPH causing the decreased ability of GSR to recycle GSSG to GSH. Compared to healthy individuals, people with T2DM have lower levels of GCLC, the rate-limiting catalytic unit in producing GSH [17]. These decreased amounts of GCLC correlated with increased amounts of transforming growth factor beta (TGF-β) in patients with T2DM [17]. Furthermore, decreased GCLC was accompanied by deceases in both GSS and gamma glutamyl transpeptidase (GGT). Both enzymes are crucial for the synthesis and import of GSH into the cell. The results show decreased overall levels of GSH in T2DM [17]. The rise in GSSG and decline of rGSH/GSH is due to the excess of free radicals or ROS being produced via the AGE/Receptor for Advanced Glycation End products (RAGE) pathway discussed in great detail below. When produced, rGSH can detoxify the ROS by accepting the electrons and combining two molecules of rGSH/GSH to make an oxidized form of glutathione GSSG.

GGT functions throughout the body as a transport molecule, enabling the breakdown of extracellular GSH. The cysteine formed is then transported into the cells for GSH biosynthesis. The mechanism by which TGF-β decreases these enzymes has not been fully elucidated; however, key components have been found. The Antioxidant Response Element (ARE) is a regulatory enhancer gene sequence, which upon activation induces the synthesis of basal and inducible GSH synthesizing enzymes. Bakin et el. found signal transduction proteins intracellularly that respond to TGF-β. Once TGF-β binds to its cellular receptor, a serine–threonine kinase, it auto-phosphorylates Smad2 and Smad3, which then heterodimerize with Smad4 and can translocate to the nucleus and upregulate or downregulate genes, such as GCLC and Glutamate cysteine ligase modifier protein (GCLM) [18]. It has also been shown that TGF-β, in addition to downregulating GCLC mRNA production, also decreases the half-life of GCLC, resulting in increasing ROS intracellularly. When TGF-β interacts with its cellular receptor, it causes an increase in caspase activity which begins the apoptosis cascade. Inducing apoptosis cleaves proteins intracellularly, including GCLC among many others [19]. With these enzymes decreased, it is recognized that people with T2DM produce a lower amount of functional GSH. Another important enzyme responsible for restoring the amount of GSH is GSR, which converts GSSG to GSH utilizing the cofactor NADPH. Contrary to the aforementioned enzymes, GSR is found in excess in patients with T2DM, which our group postulates is a compensatory mechanism to overcome GSH deficiency. However, with a deficiency in NADPH—a necessary co-factor for the activity of GSR—there is no restoration of GSH despite the increased levels of GSR. Lagman et al. also found that the amount of GSSG in T2DM was elevated in members with HbA1C over 8% [17]. They also found a five-fold decrease in Tumor Necrosis Factor alpha (TNF-α), and a two-fold decrease in IFN-γ. The study findings suggest that administering L-GSH will be more efficacious in patients with T2DM, than administering N-Acetyl Cysteine (NAC), a precursor for GSH, due to the aforementioned decreased enzymatic levels [17]. Despite abundant GSR, there are increased levels of GSSG and decreased GSH in T2DM, and this could very well be caused by the diminished levels of NADPH in lieu of polyol pathway utilization.

1.4. AGE-RAGE and GSH Deficiency in T2DM

The first AGE was discovered in the 1960s with the discovery of HbA1C. AGEs are destructive in many ways, but we will focus on one we believe to be particularly important for the function of the immune system. The binding of AGE with its counterpart Receptor for Advanced Glycation End products (RAGE) starts a cascade, increasing the generation of ROS and activation of Nuclear Factor-kappa-B (NF-$_k$B) which produces proinflammatory molecules interleukin-1 (IL-1) and interleukin-6 (IL-6). On top of the AGE-RAGE component with excess glucose in the cell, DAG is formed, which can lead to the activation of PKC. The activation of PKC involves a plethora of actions, including the activation of NF-$_k$B, further increasing IL-1 and IL-6, as well as TGF-β, which, as described previously, results in a decrease in GCLC. With the combination of TGF-β inactivating the rate-limiting step of GSH production, and with the increase in IL-1 and IL-6 there is exacerbated inflammation elevating the levels of ROS in all surrounding cells. Studies have shown that this change in the cytokine environment will lead to an altered immune response [20].

1.5. T2DM and Tuberculosis

In 2017, tuberculosis (TB) caused an estimated 1.6 million deaths and an incidence of 10 million cases of TB developed worldwide [21]. TB has been on the rise since the 1980s, in part due to the HIV pandemic, multidrug resistance, and increased numbers of highly susceptible individuals with suppressed/altered immune systems, such as those with T2DM [8]. Currently, one-quarter of the population of the world is latently infected with *M. tb* [21]. Individuals with latent tuberculosis Infection (LTBI) are not infectious; however, active TB can occur due to the reactivation of a dormant infection as a consequence of granuloma liquefaction in immunodeficient or immunocompromised individuals, such as individuals with HIV or T2DM [5]. Several studies have looked at the correlation between the prevalence of T2DM and TB. In 2012, it was found that 50% of patients studied who tested positive for *M. tb* infection had either diabetes or pre-diabetes. Other studies which found similar accounts of diabetes and TB in India found that, in Mexico, 35% of patients testing positive for *M. tb* infection had diabetes. Furthermore, in 2011, an article found that with uncontrolled diabetes and TB infection there were more cavitary lesions, and a higher incidence of positive sputum cultures, even two months after initiation treatment. Thus, careful monitoring of the progression of *M. tb* infection and glucose control should be top priorities of physicians with patients with both illnesses [22]. According to the CDC, 20% of patients with TB have diabetes in the US.

In patients with T2DM, the depletion of GSH is extremely prevalent, which may result in additional pathogenesis. Discussed later are the effects of GSH depletion in altering levels of cytokine production and the immune consequence. One cytokine that is extremely important for restricting and containing *M. tb* infection is TNF-α. This cytokine aids in the formation of granuloma to prevent the spread and wall off the pathogen. It has been shown that, in the absence of TNF-α, there will be an impairment in the formation of the granuloma, leading to the dissemination of the pathogen [23]. IFN-γ, another cytokine responsible for augmenting the effector functions of macrophages, was found to be decreased in patients with T2DM [17]. In a systematic review, patients with diabetes who were infected with *M. tb* had a risk ratio of 1.69 for treatment failure and death. The study also found an increased risk of relapse in patients with diabetes vs. non-diabetic individuals. The conclusion of the study found that improved glucose control and close monitoring in patients with the comorbidity of *M. tb* and Diabetes should be implemented to assist in the recovery process [23]. *M. tb* infection in healthy subjects will result in robust immune responses mediated by the competent immune system resulting in the formation of granuloma, which suppresses the spread of the pathogen. However, in patients with weakened immune systems, a reactivation of *M. tb* occurring in those with LTBI or primary infection can lead to active TB. We were able to find links between cytokine imbalance, as a consequence of diminished GSH levels due to decreases in the level of GSH synthesis and ability of the recycling enzymes. Increased TGF-β can be a major contributing factor to the decreasing levels of GSH synthesis enzymes. The inability of GSR in diabetic patients to restore in order to recycle GSSG to GSH is due to the depletion of NADPH by the Polyol pathway. With all these factors, it forms a sort of circle of perpetuation. (Visualized in Figure 1) *M. tb* has been a successful pathogen capable of evading the immune system, and those with a compromised immune system are at increased risk for both the reactivation of LTBI and development of active TB after initial exposure. While *M. tb* is not the only pathogen to cause an active disease in individuals with diabetes, we focused mainly on this pathogen due to the long-standing prevalence, as well as the global burden, of TB. Two cytokines—IFN- γ and TNF- α —which are imperative to the control and prevention of the spread of *M. tb*, respectively, are significantly compromised in individuals with T2DM (discussed below in the next section). We therefore believe that cytokine dysregulation in individuals with T2DM causes impaired immune responses against *M. tb*, resulting in active TB.

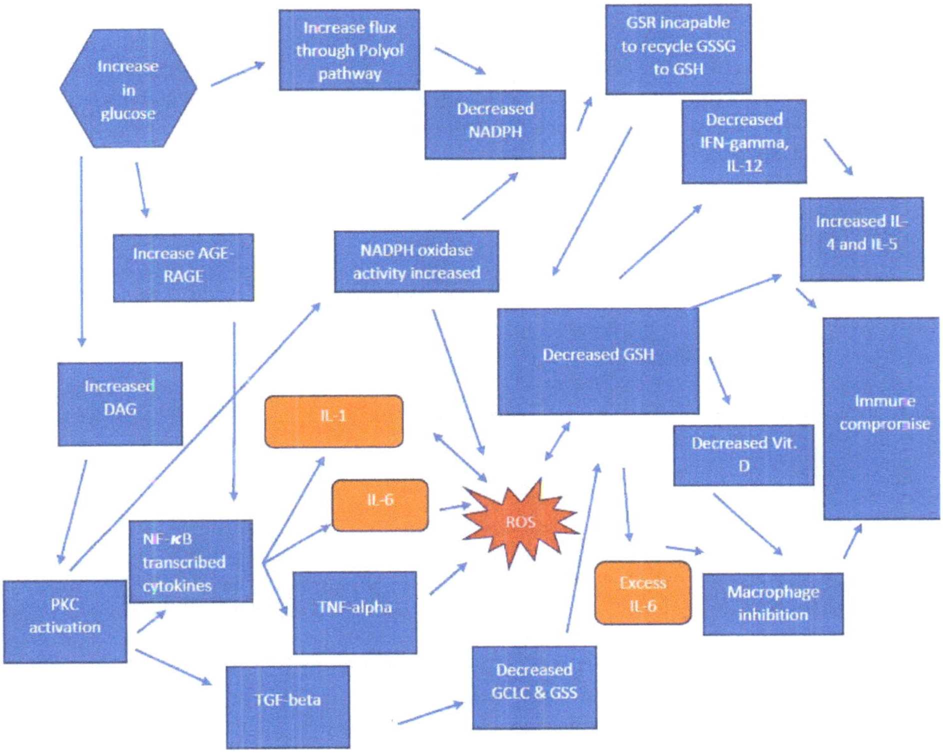

Figure 1. For oxidative stress and glutathione (GSH) deficiency in individuals with Type 2 Diabetes Mellitus (T2DM).

1.6. Cytokine Production and Immune Responses against *M. tb* Infection

Efficient communication between cells is necessary to modulate an adequate immune response in individuals for cell migration and specific instruction. This is a critical role for cytokines and chemokines. Cytokines are small, soluble proteins that are produced by the host immune cells which influence the activity of other cells by acting in a paracrine manner [24]. Different immune cells, such as macrophages and neutrophils, initiate cytokine production when interacting with a pathogen. Multiple pattern recognition receptors (PRR) on these cells recognize various bacterial factors, such as mycobacterial cell wall components, secreted molecules, and nucleic acids derived from mycobacteria, which promote cytokine production [19,20]. The cytokines discharged by these cells have essential regulatory properties, and contribute to the host defense against *M. tb* through the formation of a granuloma which leads to the containment and extermination of bacteria [25,26]. Granulomas are composed of macrophages, multinucleated giant cells, CD4+ and CD8+ T-cells, B-cells and neutrophils [27]. Within an immunocompetent individual, the interaction between *M. tb* and granulomatous cells of the immune system results in the constant secretion of cytokines; particularly, TNF-α, (formation of the granuloma), interleukin-10 (IL-10)(immune modulation), IL-6 (inflammatory cytokine), interleukin-2 (IL-2)(T-cell growth factor), interleukin-12 (IL-12)(stimulation of cells to produce IFN-γ), and IFN-γ (macrophage stimulation) [28].

TNF-α secreted by macrophages, dendritic cells (DC), and T-cells augments effector immune responses against *M. tb* infection [26,29]. Additionally, upon infection, pulmonary epithelium and lung fibroblasts produce chemokine (C-X-C motif) ligand 8 (CXCL8), which promotes the rapid recruitment of neutrophils [30]. It was previously shown, that after the ingestion of dead or dying *M. tb*-infected apoptotic neutrophils at the infection site, a proinflammatory response is triggered and macrophages become activated, releasing TNF-α [31]. Thus, it was demonstrated that TNF-α promotes the production of chemokines and chemokine receptors, and also an expression of adhesion molecules, which affect the formation of granulomas in *M. tb*-infected tissues [32]. TNF-α also works in synergy with IFN-γ to induce nitric oxide production [33].

Another important modulator of the immune response is the neutrophil-mediated release of IL-10, which is known to decrease inflammation. IL-10 is produced by phagocytes of lung lesions and reduces expression of TNF-α and IL-12 [34]. IL-10 also blocks phagosome maturation and allows the pathogen to survive in alveolar macrophages, and through reduction of T-cell responses, it may affect the integrity of the granulomas, thus promoting the transmission of *M. tb* by predisposing the host to lung cavitation [35]. IL-10 possesses macrophage-deactivating properties and reduces IL-12 production, which in turn decreases IFN-γ production by T-cells [36]. These findings indicate that IL-10 plays an anti-inflammatory role and may counter the macrophage-activating properties of IFN-γ. Therefore, the balance between TNF-α with antagonistic IL-10 and synergistic IFN-γ cytokines plays a role in fine-tuning the tissue damage in the granulomas, as well as its bactericidal effects [37].

When *M. tb* or apoptotic macrophages and neutrophils containing *M. tb* interact with DCs, it causes the DCs to mature and become able to stimulate T-cells through the expression of major histocompatibility complex (MHC) and co-stimulatory molecules, with the aid of the secretion of IL-12 and IFN-γ [38]. IL-12 inducers can polarize CD4+ T-cells toward a T-helper (T_H1) phenotype [39]. $T_H1$cells produce IFN-γ, IL-2, IL-12 and TNF-α—cytokines important for the suppression of *M. tb* growth and replication [40]. On the other hand, Type-2 T-helper phenotype (T_H2) responses lead to the production of IL-4, IL-5, and IL-10, which exert anti-inflammatory effects [41].

CD4+ and CD8+ T-cells are important in maintaining specific immunity against *M. tb*, and protective immunity is associated with antigen presentation by DC to T-cells [42]. Although MHC class II restricted CD4+ T-cells play an important role in protection against *M. tb*, MHC Class I restricted T-cells act later in the infectious cycle by killing *M. tb*-infected cells through the secretion of cytotoxic molecules, such as perforin, granzymes, and granulysin [43].

IL-2, also called T-cell growth factor, is produced by T_H1 cells. IL-2 is responsible for maintaining the CD4 and CD8 T-cell viability, and thereby amplifies T-cell responses against *M. tb* infection [37].

IL-6 is another cytokine involved in the immune response against *M. tb* which has multiple roles, including inflammation and T-cell differentiation [44]. During infection, IL-6 and TGF-β control the relative levels of expression of the transcription factor FoxP3. These T-regs have multiple inhibitory effects, and by limiting the intensity of the T-cell response to *M. tb*, they attempt to balance potentially harmful immune responses [38,39]. IL-17 mediates the cellular recruitment of neutrophils via CXCL8 in order to form lung lymphoid follicles, providing optimal macrophage activation and bacterial control during *M. tb.* infection. However, the hyperactivity of T_H17 cells can lead to an increased pathological state through the IL-17 mediated influx of neutrophils, which can cause tissue damage [41,42].

TB is a complex disease, and as such, the role of any particular cytokine cannot be categorized so easily as either "beneficial" or "harmful". However, based on the amount produced and the circumstances, these cytokines can cause either defensive or pathologic effects, by either working synergistically or antagonistically to the immune system in order to control the infection.

1.7. Why GSH Is Important in a Functioning Immune System

In 2003, Venketaraman et al. established the importance of GSH in the immune response controlling the infection of *M. tb* [45]. GSH levels were found to be decreased in red blood cells and peripheral blood mononuclear cells isolated from patients with active pulmonary TB. Dr. Venketaraman's group also found that the administration of a precursor of GSH—N-acetyl cysteine (NAC)—to isolated immune cells improved the control of *M. tb*, as well as decreased the levels of pro-inflammatory cytokines such as IL-6, TNF-α and IL-1 [45]. Furthermore, through its role as a Nitric Oxide (NO) carrier, the addition of S-Nitrosoglutathione (GSNO) to *M. tb* cultures was found to be mycobactericidal [46]. Moreover, the addition of NAC, along with cytokines IL-2, and IL-12 caused a significant enhancement in the effector functions of natural killer (NK) cells, which in turn caused an increased production of IFN-γ, a macrophage-activating cytokine [44,47–49]. The immune system requires different cytokines present to mount the proper responses. The T_H1 response, for example, which requires IL-12 and IFN-γ produced by the T_H1 cells, can further activate macrophage effector response [50]. With the infection present. ROS are produced to assist in the killing of bacteria; however, they can be harmful to the host system if not absorbed via GSH and other antioxidants to prevent cellular damage. It has been shown that ROS and GSH levels can alter the cytokine release of T-cells [51]. Depletion of GSH due to ROS production will lead to NF-$_k$B binding to the DNA, resulting in the production of IL-1, IL-6 and TNF-α. In contrast, GSH repletion will suppress the production of the aforementioned cytokines [51–53]. With a depletion of GSH during infection or T2DM, this could cause NF-$_k$B release and upregulate the production of IL-1, TNF-α, and IL-6—all inflammatory cytokines. Although IL-6 is necessary for proper macrophage activity, Van Heyningen's group found that the overproduction of IL-6 can lead to macrophages' inability to respond properly to the infection [48]. With this being said, TNF-α is an extremely important cytokine for the formation of granulomas in *M. tb* infection. A cytokine cannot be classified as "beneficial" or "detrimental"; rather, in order to produce intended effect it should be in the right site at the right time, but, as stated previously, overproduction can cause disarray, desensitization and malfunction. With each of these studies, it should now be apparent that GSH is extremely important for the proper function of the immune system, whether assisting as an NO carrier, as GSNO to assist in the killing of the infection, or scavenging excess ROS to control cytokine levels to mount a proper response. We believe the augmented immune system in diabetic patients is primarily due to the decrease in GSH, which leads to a depletion in IL-12. Deficiency of this cytokine has been shown in several studies to lead to recurrent infections, while the addition of NAC or L-GSH restored the IL-12 and IFN-γ, which in turn enhanced mycobacterial killing [44,47,54–56]. Studies have also found that when there is an increase in IL-4 and IL-5, the cytokines responsible for polarizing the helper T-cells to a T_H2 response [57,58]. Thus, further proving that the levels of ROS or GSH are a crucial deciding factor when creating an immune response.

1.8. Vitamin D and Macrophage Activation

Vitamin D is a secosteroid which can be ingested (vitamin D3 and vitamin D2) into the body or endogenously produced in the skin upon sun exposure (vitamin D3) [59,60]. Vitamin D has displayed anti-inflammatory effects, as supplementation with vitamin D results in decreased radical oxygen species (ROS) and pro-inflammatory cytokines [61,62].

GSH plays a role in maintaining vitamin D regulatory genes; GSH deficiency hinders the expression of vitamin D-binding proteins and receptors [63,64]. Furthermore, supplementation with L-cystine, a precursor for GSH, increases levels of vitamin D and its binding proteins [63,65].

Thus, we postulate that a decrease in GSH indirectly affects the immune system via decreasing the concentrations of vitamin D. However, further studies are needed to confirm at what hemoglobin A1c the deficiency of Vitamin D—due to decreases in GSH—occurs. The mechanism underlying this phenomenon can be attributed to the role of vitamin D in adaptive immunity [64]. Vitamin D assists macrophages to inhibit the proliferation of *M. tb* [59,60,66]. *M. tb* binds to the toll-like receptors seen on the cell surface of macrophages, upon which the expression of the vitamin D receptors and 1-alpha hydroxylase is upregulated [67]. The activated vitamin D then induces the transcription of human cathelicidin (hCAP18) and beta-defensin 4 (DEFB4), both of which are crucial to antimycobacterial function including the auto-lysosomal elimination of *M. tb* [59,66–69]. The importance of this intracrine pathway was demonstrated by Zhao's study in which the cellular ability to eradicate *M. tb* in DM patients improved upon vitamin D supplementation [65].

2. Summary

It has been shown in previous studies, that an excess of glucose in systemic circulation can cause an increase in ROS and the proinflammatory cytokines IL-1 and IL-6, which at high enough concentrations can inhibit the function of macrophages. Due to excess glucose, TGF-β becomes increased through the activation of AGE and its receptor RAGE, through PKC, which inhibits the manufacturing of GSH. The oxidized form of glutathione—GSSG—can be recycled if the body has enough NADPH utilizing the enzyme GSR. However, in diabetic patients, NADPH is being utilized heavily on the polyol pathway, leading to an overall decrease in GSH levels and an increase in ROS, which can alter the cytokine levels in the body [55]. GSH increase has been shown to upregulate the production of IFN-γ and IL-12, favoring the host immune response against *M. tb* infection.

The peripheral neuropathy and altered immune system can be traced back to a common metabolic pathway, but for different reasons. One of the complications of T2DM, peripheral neuropathy, was linked to the Polyol pathway and the build-up of sorbitol due to an SDH deficiency in the nervous tissue. The complications involving the immune system are not due to products from the Polyol pathway but rather from the overutilization of NADPH. The depletion of NADPH leads to the inability of GSR to recycle GSH, causing decreased GSH, increased ROS, differing cytokine profiles and, ultimately, immune system compromise.

Author Contributions: S.F., A.Y., N.N., G.L., T.N., and K.T. have contributed to drafting this review. V.V. conceived the frame work, provided guidance and assistance, and made edits to the draft.

Funding: The authors appreciate the funding support from National Heart Blood Lung Institute at the National Institutes of Health (NIH) award 1R15HL143545-01A1.

Conflicts of Interest: The authors declare no conflict of interest.

References

1. World Health Organization. *Diabetes Facts (Infographics)*; WHO: Geneva, Switzerland, 2016.
2. Centers for Disease Control and Prevention (CDC). *National Diabetes Statistics Report, 2017 Estimates of Diabetes and Its Burden in the United States Background*; CDC: Atlanta, GA, USA, 2018.
3. World Health Organization. Tuberculosis. Available online: https://www.who.int/news-room/fact-sheets/detail/tuberculosis (accessed on 13 November 2019).
4. Frieden, T.R.; Sterling, T.R.; Munsiff, S.S.; Watt, C.J.; Dye, C. Tuberculosis. *Lancet* **2003**, *362*, 887–899. [CrossRef]
5. Ai, J.-W.; Ruan, Q.-L.; Liu, Q.-H.; Zhang, W.-H. Updates on the risk factors for latent tuberculosis reactivation and their managements. *Emerg. Microbes Infect.* **2016**, *5*, 1–8. [CrossRef] [PubMed]
6. Wilcox, G. Insulin and insulin resistance. *Clin. Biochem. Rev.* **2005**, *26*, 19–39. [PubMed]
7. Nowotny, K.; Jung, T.; Höhn, A.; Weber, D.; Grune, T. Advanced Glycation End Products and Oxidative Stress in Type 2 Diabetes Mellitus. *Biomolecules* **2015**, *5*, 194–222. [CrossRef]
8. Wu, Y.; Ding, Y.; Tanaka, Y.; Zhang, W. Risk factors contributing to type 2 diabetes and recent advances in the treatment and prevention. *Int. J. Med. Sci.* **2014**, *11*, 1185–1200. [CrossRef] [PubMed]
9. Schemmel, K.E.; Padiyara, R.S.; D'Souza, J.J. Aldose reductase inhibitors in the treatment of diabetic peripheral neuropathy: A review. *J. Diabetes Complicat.* **2010**, *24*, 354–360. [CrossRef] [PubMed]
10. Oka, M.; Kato, N. Aldose Reductase Inhibitors. *J. Enzyme Inhib.* **2001**, *16*, 465–473. [CrossRef]
11. Oates, P.J. Aldose reductase, still a compelling target for diabetic neuropathy. *Curr. Drug Targets* **2008**, *9*, 14–36. [CrossRef] [PubMed]
12. Morré, D.M.; Lenaz, G.; Morré, D.J. Surface oxidase and oxidative stress propagation in aging. *J. Exp. Biol.* **2000**, *203 Pt 10*, 1513–1521.
13. Pollreisz, A.; Schmidt-Erfurth, U. Diabetic Cataract—Pathogenesis, Epidemiology and Treatment. *J. Ophthalmol.* **2010**, *2010*, 1–8. [CrossRef]
14. Cheng, H.-M.; González, R.G. The effect of high glucose and oxidative stress on lens metabolism, aldose reductase, and senile cataractogenesis. *Metabolism* **1986**, *35*, 10–14. [CrossRef]

15. Gallagher, E.J.; LeRoith, D.; Stasinopoulos, M.; Zelenko, Z.; Shiloach, J. Polyol accumulation in muscle and liver in a mouse model of type 2 diabetes. *J. Diabetes Complicat.* **2016**, *30*, 999–1007. [CrossRef] [PubMed]
16. Griffith, O.W. Biologic and pharmacologic regulation of mammalian glutathione synthesis. *Free Radic. Biol. Med.* **1999**, *27*, 922–935. [CrossRef]
17. Lagman, M.; Ly, J.; Saing, T.; Singh, M.K.; Tudela, E.V.; Morris, D.; Chi, P.-T.; Ochoa, C.; Sathananthan, A.; Venketaraman, V. Investigating the causes for decreased levels of glutathione in individuals with type II diabetes. *PLoS ONE* **2015**, *10*, e0118436. [CrossRef]
18. Bakin, A.V.; Stourman, N.V.; Sekhar, K.R.; Rinehart, C.; Yan, X.; Meredith, M.J.; Arteaga, C.L.; Freeman, M.L. Smad3-ATF3 signaling mediates TGF-β suppression of genes encoding Phase II detoxifying proteins. *Free Radic. Biol. Med.* **2005**, *38*, 375–387. [CrossRef]
19. Franklin, C.C.; Rosenfeld-Franklin, M.E.; White, C.; Kavanagh, T.J.; Fausto, N. TGFbeta1-induced suppression of glutathione antioxidant defenses in hepatocytes: Caspase-dependent post-translational and caspase-independent transcriptional regulatory mechanisms. *FASEB J.* **2003**, *17*, 1535–1537. [CrossRef]
20. Yan, Z.; Garg, S.K.; Kipnis, J.; Banerjee, R. Extracellular redox modulation by regulatory T cells. *Nat. Chem. Biol.* **2009**, *5*, 721–723. [CrossRef]
21. World Health Organization. *'Tuberculosis' Fact Sheet*; WHO: Geneva, Switzerland, 2010; Volume 104.
22. Park, S.W.; Shin, J.W.; Kim, J.Y.; Park, I.W.; Choi, B.W.; Choi, J.C.; Kim, Y.S. The effect of diabetic control status on the clinical features of pulmonary tuberculosis. *Eur. J. Clin. Microbiol. Infect. Dis.* **2012**, *31*, 1305–1310. [CrossRef]
23. Baker, M.A.; Harries, A.D.; Jeon, C.Y.; Hart, J.E.; Kapur, A.; Lönnroth, K.; Ottmani, S.-E.; Goonesekera, S.D.; Murray, M.B. The impact of diabetes on tuberculosis treatment outcomes: A systematic review. *BMC Med.* **2011**, *9*, 81. [CrossRef]
24. Dinarello, C.A. Historical insights into cytokines. *Eur. J. Immunol.* **2007**, *37* (Suppl. 1), 34–45. [CrossRef]
25. Domingo-Gonzalez, R.; Prince, O.; Cooper, A. Cytokines and Chemokines in Mycobacterium tuberculosis Infection. In *Tuberculosis and the Tubercle Bacillus*, 2nd ed.; American Society of Microbiology: Washington, DC, USA, 2016; pp. 33–72.
26. Kaufmann, S.H.E. Protection against tuberculosis: Cytokines, T cells, and macrophages. *Ann. Rheum. Dis.* **2002**, *61* (Suppl. 2), ii54–ii58. [CrossRef] [PubMed]
27. Lin, P.L.; Plessner, H.L.; Voitenok, N.N.; Flynn, J.L. Tumor Necrosis Factor and Tuberculosis. *J. Investig. Dermatol. Symp. Proc.* **2007**, *12*, 22–25. [CrossRef] [PubMed]
28. Cooper, A.M.; Mayer-Barber, K.D.; Sher, A. Role of innate cytokines in mycobacterial infection. *Mucosal Immunol.* **2011**, *4*, 252–260. [CrossRef] [PubMed]
29. Philips, J.A.; Ernst, J.D. Tuberculosis Pathogenesis and Immunity. *Annu. Rev. Pathol. Mech. Dis.* **2012**, *7*, 353–384. [CrossRef] [PubMed]
30. Wickremasinghe, M.I.; Thomas, L.H.; Friedland, J.S. Pulmonary epithelial cells are a source of IL-8 in the response to Mycobacterium tuberculosis: Essential role of IL-1 from infected monocytes in a NF-κB-dependent network. *J. Immunol.* **1999**, *163*, 3936–3947. [PubMed]
31. Alexander, Y.; Persson, Z.; Blomgran-Julinder, R.; Rahman, S.; Zheng, L.; Stendahl, O. Mycobacterium tuberculosis-induced apoptotic neutrophils trigger a pro-inflammatory response in macrophages through release of heat shock protein 72, acting in synergy with the bacteria. *Microbes Infect.* **2008**, *10*, 233–240.
32. Flynn, J.L.; Chan, J. I Mmunology of T Uberculosis. *Annu. Rev. Immunol.* **2001**, *19*, 93–129. [CrossRef]
33. Liew, F.Y.; Li, Y.; Millott, S. Tumor necrosis factor-alpha synergizes with IFN-gamma in mediating killing of Leishmania major through the induction of nitric oxide. *J. Immunol.* **1990**, *145*, 4306–4310.
34. Beamer, G.L.; Flaherty, D.K.; Assogba, B.D.; Stromberg, P.; Gonzalez-Juarrero, M.; de Waal Malefyt, R.; Vesosky, B.; Turner, J. Interleukin-10 promotes Mycobacterium tuberculosis disease progression in CBA/J mice. *J. Immunol.* **2008**, *181*, 5545–5550. [CrossRef]
35. O'Leary, S.; O'Sullivan, M.P.; Keane, J. IL-10 Blocks Phagosome Maturation in *Mycobacterium tuberculosis*–Infected Human Macrophages. *Am. J. Respir. Cell Mol. Biol.* **2011**, *45*, 172–180. [CrossRef]
36. Gong, J.H.; Zhang, M.; Modlin, R.L.; Linsley, P.S.; Iyer, D.; Lin, Y.; Barnes, P.F. Interleukin-10 downregulates Mycobacterium tuberculosis-induced Th1 responses and CTLA-4 expression. *Infect. Immun.* **1996**, *64*, 913–918. [PubMed]
37. Lowe, D.M.; Redford, P.S.; Wilkinson, R.J.; O'Garra, A.; Martineau, A.R. Neutrophils in tuberculosis: Friend or foe? *Trends Immunol.* **2012**, *33*, 14–25. [CrossRef] [PubMed]

38. Giacomini, E.; Iona, E.; Ferroni, L.; Miettinen, M.; Fattorini, L.; Orefici, G.; Julkunen, I.; Coccia, E.M. Infection of Human Macrophages and Dendritic Cells with *Mycobacterium tuberculosis* Induces a Differential Cytokine Gene Expression That Modulates T Cell Response. *J. Immunol.* **2001**, *166*, 7033–7041. [CrossRef] [PubMed]
39. Sano, K.; Haneda, K.; Tamura, G.; Shirato, K. Ovalbumin (OVA) and *Mycobacterium tuberculosis* Bacilli Cooperatively Polarize Anti-OVA T-helper (Th) Cells toward a Th1-Dominant Phenotype and Ameliorate Murine Tracheal Eosinophilia. *Am. J. Respir. Cell Mol. Biol.* **1999**, *20*, 1260–1267. [CrossRef]
40. Boom, W.H.; Canaday, D.H.; Fulton, S.A.; Gehring, A.J.; Rojas, R.E.; Torres, M. Human immunity to M. tuberculosis: T cell subsets and antigen processing. *Tuberculosis (Edinb)* **2003**, *83*, 98–106. [CrossRef]
41. Mosmann, T.R.; Coffman, R.L. TH1 and TH2 Cells: Different Patterns of Lymphokine Secretion Lead to Different Functional Properties. *Annu. Rev. Immunol.* **1989**, *7*, 145–173. [CrossRef] [PubMed]
42. Cooper, A.M. T cells in mycobacterial infection and disease. *Curr. Opin. Immunol.* **2009**, *21*, 378–384. [CrossRef] [PubMed]
43. Ottenhoff, T.H.M.; Kaufmann, S.H.E. Vaccines against Tuberculosis: Where Are We and Where Do We Need to Go? *PLoS Pathog.* **2012**, *8*, e1002607. [CrossRef]
44. VanHeyningen, T.K.; Collins, H.L.; Russell, D.G. IL-6 produced by macrophages infected with Mycobacterium species suppresses T cell responses. *J. Immunol.* **1997**, *158*, 330–337.
45. Venketaraman, V.; Millman, A.; Salman, M.; Swaminathan, S.; Goetz, M.; Lardizabal, A.; Hom, D.; Connell, N.D. Glutathione levels and immune responses in tuberculosis patients. *Microb. Pathog.* **2008**, *44*, 255–261. [CrossRef]
46. Venketaraman, V.; Dayaram, Y.K.; Talaue, M.T.; Connell, N.D. Glutathione and Nitrosoglutathione in Macrophage Defense against Mycobacterium tuberculosis. *Infect. Immun.* **2005**, *73*, 1886–1889. [CrossRef] [PubMed]
47. Afzali, B.; Mitchell, P.; Lechler, R.I.; John, S.; Lombardi, G. Translational Mini-Review Series on Th17 Cells: Induction of interleukin-17 production by regulatory T cells. *Clin. Exp. Immunol.* **2010**, *159*, 120–130. [CrossRef] [PubMed]
48. Millman, A.C.; Salman, M.; Dayaram, Y.K.; Connell, N.D.; Venketaraman, V. Natural Killer Cells, Glutathione, Cytokines, and Innate Immunity Against *Mycobacterium tuberculosis*. *J. Interf. Cytokine Res.* **2008**, *28*, 153–165. [CrossRef] [PubMed]
49. Kasai, M.; Yoneda, T.; Habu, S.; Maruyama, Y.; Okumura, K.; Tokunaga, T. In vivo effect of anti-asialo GM1 antibody on natural killer activity. *Nature* **1981**, *291*, 334–335. [CrossRef]
50. Janeway, C.A., Jr.; Travers, P.; Walport, M. *Immunobiology: The Immune System in Health and Disease*, 5th ed.; Garland Science: New York, NY, USA, 2001.
51. Garg, S.K.; Yan, Z.; Vitvitsky, V.; Banerjee, R. Differential dependence on cysteine from transsulfuration versus transport during T cell activation. *Antioxid. Redox Signal.* **2011**, *15*, 39–47. [CrossRef]
52. Verhasselt, V.; Berghe, W.V.; Vanderheyde, N.; Willems, F.; Haegeman, G.; Goldman, M. N-acetyl-L-cysteine inhibits primary human T cell responses at the dendritic cell level: Association with NF-kappaB inhibition. *J. Immunol.* **1999**, *162*, 2569–2574.
53. Bernal-Fernandez, G.; Espinosa-Cueto, P.; Leyva-Meza, R.; Mancilla, N.; Mancilla, R. Decreased Expression of T-Cell Costimulatory Molecule CD28 on CD4 and CD8 T Cells of Mexican Patients with Pulmonary Tuberculosis. *Tuberc. Res. Treat.* **2010**, *2010*, 1–8. [CrossRef]
54. Haraguchi, S.; Day, N.K.; Nelson, R.P.; Emmanuel, P.; Duplantier, J.E.; Christodoulou, C.S.; Good, R.A. Interleukin 12 deficiency associated with recurrent infections. *Proc. Natl. Acad. Sci. USA* **1998**, *95*, 13125–13129. [CrossRef]
55. Trinchieri, G. Interleukin-12 and the regulation of innate resistance and adaptive immunity. *Nat. Rev. Immunol.* **2003**, *3*, 133–146. [CrossRef]
56. Tan, K.S.; Lee, K.O.; Low, K.C.; Gamage, A.M.; Liu, Y.; Tan, G.Y.G.; Koh, H.Q.V.; Alonso, S.; Gan, Y.H. Glutathione deficiency in type 2 diabetes impairs cytokine responses and control of intracellular bacteria. *J. Clin. Investig.* **2012**, *122*, 2289–2300. [CrossRef]
57. Kumar, N.P.; Sridhar, R.; Banurekha, V.V.; Jawahar, M.S.; Fay, M.P.; Nutman, T.B.; Babu, S. Type 2 diabetes mellitus coincident with pulmonary tuberculosis is associated with heightened systemic type 1, type 17, and other proinflammatory cytokines. *Ann. Am. Thorac. Soc.* **2013**, *10*, 441–449. [CrossRef] [PubMed]
58. Chen, H.; Wen, F.; Zhang, X.; Su, S.B. Expression of T-helper-associated cytokines in patients with type 2 diabetes mellitus with retinopathy. *Mol. Vis.* **2012**, *18*, 219–226. [PubMed]

59. Sutaria, N.; Liu, C.-T.; Chen, T.C. Vitamin D Status, Receptor Gene Polymorphisms, and Supplementation on Tuberculosis: A Systematic Review of Case-Control Studies and Randomized Controlled Trials. *J. Clin. Transl. Endocrinol.* **2014**, *1*, 151–160. [CrossRef] [PubMed]
60. Zhao, X.; Yuan, Y.; Lin, Y.; Zhang, T.; Bai, Y.; Kang, D.; Li, X.; Kang, W.; Dlodlo, R.A.; Harries, A.D. Vitamin D status of tuberculosis patients with diabetes mellitus in different economic areas and associated factors in China. *PLoS ONE* **2018**, *13*, e0206372. [CrossRef]
61. Jain, S.K.; Micinski, D. Vitamin D upregulates glutamate cysteine ligase and glutathione reductase, and GSH formation, and decreases ROS and MCP-1 and IL-8 secretion in high-glucose exposed U937 monocytes. *Biochem. Biophys. Res. Commun.* **2013**, *437*, 7–11. [CrossRef]
62. Farrokhian, A.; Raygan, F.; Bahmani, F.; Talari, H.R.; Esfandiari, R.; Esmaillzadeh, A.; Asemi, Z. Long-Term Vitamin D Supplementation Affects Metabolic Status in Vitamin D–Deficient Type 2 Diabetic Patients with Coronary Artery Disease. *J. Nutr.* **2017**, *147*, 384–389. [CrossRef]
63. Parsanathan, R.; Jain, S.K. Glutathione deficiency alters the vitamin D-metabolizing enzymes CYP27B1 and CYP24A1 in human renal proximal tubule epithelial cells and kidney of HFD-fed mice. *Free Radic. Biol. Med.* **2019**, *131*, 376–381. [CrossRef]
64. Jain, S.K.; Parsanathan, R.; Achari, A.E.; Kanikarla-Marie, P.; Bocchini, J.A. Glutathione Stimulates Vitamin D Regulatory and Glucose-Metabolism Genes, Lowers Oxidative Stress and Inflammation, and Increases 25-Hydroxy-Vitamin D Levels in Blood: A Novel Approach to Treat 25-Hydroxyvitamin D Deficiency. *Antioxidants Redox Signal.* **2018**, *29*, 1792–1807. [CrossRef]
65. Jain, S.K.; Kahlon, G.; Bass, P.; Levine, S.N.; Warden, C. Can l-cysteine and Vitamin D rescue Vitamin D and Vitamin D binding protein levels in blood Plasma of African American type 2 diabetic patients? *Antioxid. Redox Signal.* **2015**, *23*, 688–693. [CrossRef]
66. Chesdachai, S.; Zughaier, S.M.; Hao, L.; Kempker, R.R.; Blumberg, H.M.; Ziegler, T.R.; Tangpricha, V. The effects of first-line anti-tuberculosis drugs on the actions of vitamin D in human macrophages. *J. Clin. Transl. Endocrinol.* **2016**, *6*, 23–29. [CrossRef]
67. Hewison, M. Vitamin D and the intracrinology of innate immunity. *Mol. Cell. Endocrinol.* **2010**, *321*, 103–111. [CrossRef] [PubMed]
68. Ashenafi, S.; Mazurek, J.; Rehn, A.; Lemma, B.; Aderaye, G.; Bekele, A.; Assefa, G.; Chanyalew, M.; Aseffa, A.; Andersson, J.; et al. Vitamin D3 status and the association with human cathelicidin expression in patients with different clinical forms of active tuberculosis. *Nutrients* **2018**, *10*, 721. [CrossRef] [PubMed]
69. Herrera, M.T. Gonzalez, Y.; Hernández-Sánchez, F.; Miguel, G.F.; Torres, M. Low serum vitamin D levels in type 2 diabetes patients are associated with decreased mycobacterial activity. *BMC Infect. Dis.* **2017**, *17*, 1–9. [CrossRef] [PubMed]

Journal of
Clinical Medicine

MDPI

Review

Potentials of Host-Directed Therapies in Tuberculosis Management

Yash Dara †, Doron Volcani †, Kush Shah †, Kevin Shin † and Vishwanath Venketaraman *

Department of Basic Medical Sciences, College of Osteopathic Medicine of the Pacific, Western University of Health Sciences, Pomona, CA 91766, USA

* Correspondence: vvenketaraman@westernu.edu; Tel.: +1-909-706-3736

† All of these authors contributed equally to the presented review.

Received: 27 June 2019; Accepted: 2 August 2019; Published: 3 August 2019

Abstract: Tuberculosis (TB) remains as a leading cause of mortality in developing countries, persisting as a major threat to the global public health. Current treatment involving a long antibiotic regimen brings concern to the topic of patient compliance, contributing to the emergence of drug resistant TB. The current review will provide an updated outlook on novel anti-TB therapies that can be given as adjunctive agents to current anti-TB treatments, with a particular focus on modulating the host immune response to effectively target all forms of TB. Additional potential therapeutic pathway targets, including lipid metabolism alteration and vascular endothelial growth factor (VEGF)-directed therapies, are discussed.

Keywords: *Mycobacterium tuberculosis*; host-directed therapies; autophagy; immune responses

1. Introduction

Tuberculosis (TB), caused by the bacterial pathogen *Mycobacterium tuberculosis* (*M. tb*), remains a leading cause of mortality in developing countries, with approximately one-third of the global population remaining infected and 10 million new cases reported officially to the World Health Organization (WHO) in 2017 [1]. Currently, the best first-line curative therapy strategy for TB is the Directly Observed Treatment, Short Course (DOTS), comprising of an antibiotic regimen of isoniazid, rifampicin, pyrazinamide, and ethambutol for six to nine months, and additional patient monitoring during the first two initial months. However, due to its long duration and adverse side effects, patient compliance greatly decreases, leading to discontinuation of DOTS. This has contributed to the emergence of multidrug-resistant TB (MDR-TB), representing the continuous threat it possesses to public health and economic growth, globally [2]. As antimicrobial resistance against *M. tb* increases, it has become an international priority to develop effective, novel therapeutic approaches.

Recent improvements in the field of genomics have provided insight into various pathways that can be exploited as targets of possible emerging therapies. Along with the relatively successful treatment methodology of the pre-chemotherapy era, which demonstrated a capacity of 'self-cure' against TB, a shift in focus towards modulating the host immune response as adjunctive therapy, namely host-directed therapies (HDTs), to meet the needs for nearly all forms of *M. tb* infections, including MDR-TB, is underway. The emerging and existing HDT agents and their potentials will be the subject of the current review.

2. mTOR Inhibition

In recent years, there is an interest in exploiting autophagy pathways via HDTs as an approach to manage TB [3,4]. This cellular process, involving lysosomal degradation, is essential in the removal of protein aggregates and damaged cellular organelles that may damage the integrity of the cell [3]. A summary of the autophagy pathway is described in the following: as autophagy begins, a double-membrane bound vesicle containing cytoplasmic material, namely an autophagosome, is formed. These autophagosomes will fuse with lysosomes to form autophagolyosomes, resulting in degradation of cellular debris [4]. While this process is complex, there are three required components, Unc-51-like Kinase 1 complex (ULK1), focal adhesion kinase family interacting protein of 200 kd (FIP200), and autophagy related protein 13 (ATG13) [4].

Mammalian target of rapamycin (mTOR) is a multi-process regulator, specifically involved in numerous anabolic pathways, thus being most active when nutrients are readily available, aiding in cell survivability [4,5]. Typically, mTOR is inactivated in conditions of cellular stress (i.e., hypoxia and starvation), inducing autophagy. One proposed mechanism of mTOR interaction in autophagy via phosphorylation of ATG13 to inhibit the actions of the ULK complex [5]. Additionally, another discussed mechanism is the inhibition of the active form of Beclin-1-regulated autophagy protein [5]. Downstream metabolic effects after mTOR activation have been previously described: mTOR-1 complex activates SREBP-1 and PPAR-gamma to promote lipogenesis and activates S6K1 to promote glucose transporter 1 (GLUT-1) synthesis to increase glycolysis and lipogenesis. Additionally, mTOR-1 complex activates S6K1 to promote protein synthesis [5].

In the case of TB, the key role of autophagy is to restrict *M. tb* growth and can be triggered to do so via numerous intracellular cues (i.e., IFN-gamma and vitamin D) [5]. However, it has been observed that in the presence of *M. tb*, mTOR activity is continuously expressed, and thus, will promote the aerobic glycolytic pathway [5]. This shift in metabolism, being similar to the Warburg effect in cancer cells, is thought to be essential in mounting an immune response against *M. tb* [6]. Because mTOR is a negative regulator of autophagy, it can be noted that *M. tb* survivability is correlated with the inhibition of autophagolysosome fusion [6]. As such, stimulating autophagy via various pathways will increase the ability to fight *M. tb* infection [5–7]. One of the main strategies to tackle *M. tb* is through direct inhibition of the mTOR via rapamycin analogs, with greater focus on everolimus due to its greater bioavailability and lower side effect profile [7]. While the beneficial effects of everolimus has been demonstrated over the years, it is important to note high doses of everolimus is also an effective immunosuppressant and has applications in cases of organ transplants and cancer therapies. There have been discussions of utilizing various drug delivery systems to target *M. tb* specific cell [4].

Looking at in vivo effects of rapamycin and everolimus reiterates the potential yet further investigation needed when using mTOR inhibition for treating *M. tb* infection. In a Zebrafish model with *M. marinum* infection, mTOR-deficient Zebrafish cleared infection earlier [8]. In mice models, rapamycin given to mice at an early age did not significantly change life expectancy or susceptibility to disease; however, when given at a later age, the mice had better survival expectancy [9]. When rapamycin was given to BCG-vaccinated mice, there was an increase in the vaccination efficacy against *M. tb* infection associated with autophagy, increased antigen presentation, and increased Th1-type immune response [10]. However, when investigating rapamycin efficiency in cells pre-infected with HIV, the mTOR inhibition was advantageous for *M. tb* [11]. The rapamycin derivative, everolimus, also shows promise in vivo. In healthy elderly volunteers given low doses of everolimus, there was a 20% improvement in protection after the influenza vaccination, with low doses showing the lowest number of adverse events [12]. In contrast, when looking at organ-transplant patients given everolimus, there was a higher risk of TB and reactivation of latent TB [13]. The doses in this study were higher than that of the influenza study, and the higher risk associated with higher doses of everolimus could be overcome with Rifampicin [14]. Overall, more needs to be done to understand the dose-response of everolimus. Alternatively, enhancing the delivery of such drugs could have better potential against *M. tb*. For example, an in vitro study looking at inhalable rapamycin in particles showed "macrophages exposed to the particles or rapamycin in solution at a concentration of 100 μg/mL over a 24 h period maintained 79.37 ± 0.72% and 58.33 ± 1.39% viability, respectively" [15]. They further concluded inhalable rapamycin to be better at clearing *M. tb* than rapamycin in solution in vitro [15].

3. Cathelicidin (LL37) Inducers

Cathelicidins are part of the antimicrobial peptide class, with LL37 being the only known human cathelicidin [16]. They kill mycobacteria, regulate the innate immune system through cytokines and chemokines, and participate in autophagy [17]. LL37 is 37 amino acids long, alpha-helical, and cationic making it have a higher affinity to DNA, and aggregates in solution [16]. It penetrates the bacterial membrane, causing pores and bacterial lysis, but it does not affect mammalian cells with cholesterol in their membranes [16]. LL37 is described to have both proinflammatory and anti-inflammatory effects depending on its environment [16]. Cathelicidins are expressed in neutrophils, monocytes, keratinocytes, lymphocytes, and epithelial cells of the skin, testis, gastrointestinal system, and respiratory tract [17].

LL37 produced by alveolar macrophages and neutrophils significantly contributes to the growth inhibition of intracellular pathogens [17]. Specifically, *M. tb* is shown to induce synthesis of LL37 in epithelial cells, neutrophils, monocyte-derived macrophages, and in alveolar macrophages [17]. However, LL37 levels were undetectable in TB granulomas, indicative of its absence during late stages of *M. tb* infection. *M. tb* increases the amount of LL37 by activating Toll-Like Receptors (TLRs) 2, and especially 9 [18]. In this proinflammatory pathway, the activated TLR9 will induce the synthesis of Type I IFN, LL37 and thereby forming M1 macrophages, NETosis (neutrophil extracellular traps) with DNA complexes, and increased inflammatory cells [16]. When LL37 produced during *M. tb* infection triggers formation of NETs, the NET: LL37complexes containing *M. tb* will then be internalized by the macrophages, followed by the killing of the pathogen inside the lysosomes [19].

The importance of cathelicidins and their protection against TB are seen in the animal studies below [20,21]. It has been shown that mycobacterial infection can increase the levels of cAMP, which inhibits cathelicidins and therefore promotes intracellular *M. tb* growth. This was demonstrated in mice lacking the cathelicidin-producing gene. This study used Cramp-knocked out mice (the gene equivalent to human LL37 gene), to see if it was required for regulating protective immunity against *M. tb* in vivo. They describe the experiment as using "Cramp−/− mice in a validated model of pulmonary tuberculosis and conducted cell-based assays with macrophages from these mice" [20]. The results showed "macrophages from Cramp−/− mice were unable to control M. tuberculosis growth in an in vitro infection model, were deficient in intracellular calcium influx and were defective in stimulating T-cells [20]. Additionally, CD4 and CD8 T-cells from Cramp−/− mice produced less IFNβ upon stimulation. Furthermore, bacterial-derived cyclic-AMP modulated cathelicidin expression in macrophages" [20]. Thus, it is necessary to have cathelicidins for the innate response to protect against *M. tb* infections [20]. In another study, treatment of immunocompromised mice that are latently infected with *M. tb* with combination of TNF-alpha, beta-defensin, and LL37 resulted in protection and prevented reactivation of latent *M. tb* infection. This further confirms the ability of LL37 in protecting against reactivation of latent *M. tb* infection. Still, there is some caution needed when considering LL37 as a therapeutic, as it has been shown that in people who were resistance to Colistin (an antibiotic usually given as a last line of defense to drug resistant bacteria), were also resistant to LL37 [21].

Recent studies have demonstrated the potential of vitamin D administration to induce LL37 production via activation of TLR2/1 on human macrophages [22,23]. Vitamin D causes increase in cathelicidin release, which upregulates transcription of Beclin-1 and Atg5, which help in autophagy [24]. It is important to note that without LL37, vitamin D does not decrease *M. tb* growth, as demonstrated by using siRNA to knockout LL37 gene [25]. In addition, Vitamin D may also inhibit *M. tb* growth by enhancing the ability of monocytes to respond to IFN-gamma [26]. It then seems promising to use Vitamin D supplementation for individuals more susceptible to *M. tb* infection, such as HIV patients [27]. However, there are conflicting reports regarding the effectiveness of serum-25 hydroxyvitamin D in cases of active pulmonary tuberculosis. In one cross-sectional study, it was determined that HIV patients, testing positive for pulmonary TB, still had greater serum levels of 25-hydroxyvitamin D [27]. While this is at variance with other reports, there are still clinical trials demonstrating beneficial potential of adjunctive vitamin D supplementation. Since vitamin D, as well as 4-phenylbutyrate

(PBA), have been shown to induce LL37 release, one study found giving either vitamin D, PBA, or both in conjunction with chemotherapy to adults with active TB provided beneficial effects towards clinical recovery [28]. Currently, there is an ongoing clinical trial looking at the effectiveness of vitamin D supplementation in adjunct to antiretroviral therapy for HIV-infected adults with low serum-25 hydroxyvitamin D since these individuals are at higher risk for mortality, HIV progress, and incidence of TB [29]. This study's findings are expected to lead to larger vitamin D supplementation trials for preventing pulmonary TB in other populations as well [29].

4. Adjunctive Defensin Therapy

Defensins are another group of antimicrobial peptides that are part of the innate immune system [30]. They are arginine rich, cationic, 16–50 amino acids long, that form beta-sheets held by three disulfide bonds. Defensins have pharmaceutical potential due to their low molecular weight, non-specific and broad activity, and are resistant to proteolysis. There are three subfamilies: alpha, beta, and theta. They differ based on their length, disulfide bond locations, and structure of their precursor. Alpha and beta are more similar as compared to theta in amino acid sequence, tertiary structures, and location of their genes on the same chromosome [30]. Defensins are stored in the granules and lysosomes of innate immune and are released into the environment. They help against many pathogens, including gram positive and negative bacteria, mycobacteria, fungi, and viruses [30].

There are six subtypes of human alpha defensins. They are stored in neutrophils, as well as tracheal epithelial cells, saliva, mucosa of the cheeks, submandibular glands, and the Paneth cells of the small intestines [30]. Alpha defensins work against gram positive and negative bacteria, mycobacteria, fungi, and viruses. There are 11 subtypes of human beta defensins, which are seen in the intestines, reproductive system, epithelial cells of the trachea and bronchi, mucous, and macrophages [30]. They have antimicrobial effects on gram positive and negative bacteria, multi-resistant bacteria, and yeast. As for the theta defensins, they have not been found in humans, and while they have shown antimicrobial activity, they have not been shown to be effective against mycobacteria [30].

Besides their antimicrobial effects, defensins also increase histamine release by activating mast cells, increase cytokines, act as ligands, help in wound healing, and beta defensins increase bone growth [30].

More specifically, *M. tb* infection is frequently associated with the production of human beta defensin-2 (HBD-2) [31]. It has been shown that *M. tb* induces HBD-2 mRNA expression in lung epithelial cells, and alveolar macrophages. HBD-2 peptides are seen in lung epithelial cells and are more concentrated where there are *M. tb* clusters [31]. During progressive pulmonary TB, there is an initial rise in HBD-2, which inhibits bacilli proliferation. During latent TB, there is a continuous increase in HBD-2, however when TB is reactivated, there is low amounts of defensin [18]. There are also correlations with alpha defensins and TB, as seen when comparing patients with TB and healthy individuals who previously had TB both showing an increase in alpha 1, 3, and 4 defensins, seen in eosinophils [17].

More specific to the subtypes of beta defensins, HBD-1 is constitutively released, whereas 2-4 are induced by pathogens. HBD-1 and 2 are important for permeabilizing *M. tb.*, and HBD-2 is especially seen in the respiratory tract where they are induced by bacteria and cytokines, go to macrophages or phagosomes, and are activated when LPS reacts with TLR4 [17,32]. HBD-4 has shown to be released in infected macrophages by TLR2/1 stimulation induced by IL-1B and vitamin D [17,32]. L-isoleucine has also shown to upregulate gene expression of HBD-3 and 4, which lowers the bacilli load [32].

One mechanism as to how defensins work was shown *in vitro*, where they used 50 mg/mL to inhibit *M. tb* growth, independent of calcium and magnesium [33]. Lesions were seen on the surface of the bacteria [33]. This means there is an increase in permeabilizing the cell membranes of the bacteria, which was further shown to occur extracellularly or during phagocytosis. Defensins are also attracted to glycoproteins to kill viruses with envelopes, and polyanionic structures such as DNA. The cationic properties of defensins also enable them to be attracted towards the anionic phospholipids of bacteria and viruses, in addition to DNA [32].

Another mechanism of defensins is their chemotactic effects [32]. Alpha defensins recruit macrophages, while beta defensins recruit immature T-cells and dendritic cells through CCR6 receptors. Theta defensins have no chemotactic effects [32]. *M. tb* causes macrophages to release HBD-2 through TLR and recognizing pathogen-associated molecular patterns such as peptidoglycan and lipoteichoic acid. This is done through MAPK or NF-kB, which further causes proinflammation and cytokines to be released [32].

Looking from a future pharmaceutical and therapeutic perspective, some of the challenges include the excessive cost, increase in concentration causing cytotoxicity, and being less stable in vivo [30,32]. Still, there have been positive signs in using defensins at much lower concentrations in vivo. In this study, mice infected with *M. tb* needed 5 μg of HNP-1 given subcutaneously to have a significant decrease in CFU from the lungs in a matter of 1 week. Even further, "therapy with 1 μg of HNP-1 also resulted in a significant decrease in CFU from lungs after 2 and 4 weeks ($P < 0.01$ for both time points) compared to controls" [34]. Alternatively, another in vivo study successfully used L-isoleucine to induce beta-defensins. Mice were infected with *M. tb* for 60 days were then given 250 μg of intratracheal L-isoleucine every 48 h. The results showed "administration of l-isoleucine induced a significant increase of beta-defensins 3 and 4 which was associated with decreased bacillary loads and tissue damage" [35]. However, another challenge to be aware of is how bacteria resistant to antibiotics with the *mprF* gene seem also to be resistant to antimicrobial peptides. This is because that gene makes the pathogen less negatively charged. However, because defensins are non-specific, there is overall fewer resistant bacteria to this anti-microbial peptide (AMP) compared to antibiotics [32]. There have been few reports of administering defensins as therapy. Most promising is their synergistic effects with anti-Tb drugs, by increasing the permeability of the bacteria, allowing the drugs to interact with its intracellular target more easily [32]. More research should investigate the effects of L-isoleucine and vitamin D as well.

5. Metformin

Metformin, a widely known diabetes drug, is one of the front-line medications in the management of type 2 diabetes. Metformin promotes autophagy by altering the AKT-mTOR signaling pathway [36]. It increases phosphorylation of AMPK and decreases the production of various cytokines [36]. Experimental TB mice model studies demonstrated the potential benefits metformin has in improving pulmonary pathology and reducing bacterial load. However, modest effects were shown when metformin was used as an adjunctive therapy with isoniazid, a common anti-TB drug [37]. With recent cohort studies performed in India and Taiwan, there is increasing evidence suggesting metformin usage being associated with a lower risk of incident TB [38,39]. Further clinical research is needed in thoroughly understanding the role of metformin in diabetic patients infected with TB.

6. Statins

Altering lipid metabolism via statin therapy is another potential target for novel TB management therapies [7]. Statins (HMG-CoA reductase inhibitors) are utilized to lower serum LDL cholesterol to prevent issues like atherosclerosis. In addition, statins are also known to have anti-inflammatory effects mediated by transforming growth factor-beta and peroxisome proliferator-activated receptor-gamma [40]. In the case of TB, statins promote phagosome maturation and induce autophagy through reductions in cholesterol levels since mycobacteria preferentially utilizes lipid carbon sources as nutrients [41]. Experimental studies using the statin simvastatin as an adjunctive therapy in conjunction with rifampicin, isoniazid, and pyrazinamide resulted in significant reduction in the bacterial burden [42]. Despite being very promising, a large retrospective analysis of statin usage for diabetic patients infected with *M. tb* were not associated with a protective effect [43]. Therefore, it is essential to perform additional research in this area to determine optimal statin agents, including repurposing current FDA-approved drugs, with appropriate dosing schedules for the most effective results.

7. Additional Approaches

Additional emerging approaches that have not been discussed extensively in the literature include nanoparticles, vitamin A supplementation, and anti-VEGF inhibitors.

Nanoparticles show high promise as an innovative drug delivery platform by providing a controlled environment in which various drugs can be continuously released to improve treatment outcomes [44]. Numerous studies have demonstrated enhanced intracellular accumulation of three anti-TB drugs, namely isoniazid, rifampin, and streptomycin [44]. Antimicrobial activity of isoniazid and streptomycin was enhanced, while there was no notable change in rifampin activity while encapsulated in nanoparticle [44]. As improvements in stability of these nanoparticles system progress, it is highly possible that novel adjunctive therapies can be administered via nanoparticle encapsulation [44]. The other potential with using nanoparticles is lowering cytotoxicity, as demonstrated in a Zebrafish model of tuberculosis study, when thioridazine was encapsulated in a nanoparticle, no toxicity was detected compared to free thioridazine. This study also found the thioridazine nanoparticle therapy improved rifampicin's effect against *M. tb* in vivo [45].

Vitamin A deficiency has been observed in patients with TB. The deficiency of vitamin A in patients with TB might be a contributor to the development of TB [46]. Alternatively, deficiency could be the result of loss of appetite, poor intestinal absorption, increased urinary loss of vitamin A due to TB [46]. Vitamin A deficiency lowers immunity whilst vitamin A supplementation reduces morbidity and mortality, particularly from measles and diarrhea. Vitamin A supplementation also decreases the mortality rate in HIV-infected children and delays the progression to AIDS in HIV infected subjects [46]. A higher incidence of lung cancer and increased mortality have been observed in smokers after beta-carotene supplementation. It is thought that multiple micronutrients rather than vitamin A alone may be more beneficial in patients with TB. However, further research is needed to be conclusive.

Since TB granulomas feature abnormal vasculature, there is discussion of administering anti-TB drugs in conjunction with antiangiogenic (anti-VEGF) agents, such as bevacizumab [47]. This has the potential to improve delivery of antimicrobial drugs into granulomas and stimulate sensitivity to drug treatment through improved oxygen delivery into the lesions during the window of normalization [47]. Furthermore, this type of 'host-directed therapy' that targets the abnormal granuloma vasculature and reduces hypoxia may lead to a more robust immune response again the bacteria. Through this mechanism, vessel normalization has the potential to reduce the overall duration of TB chemotherapies and possibly avoid localized exposure of the bacterium to monotherapy, thereby avoiding the development of drug resistance, which is one of the main issues in the global control of TB [47].

8. Conclusions

A current review of the developing HDT agents presents as a challenge due to the broad spectrum of research within the field. Current data, both pre- and clinical, is insufficient to draw major conclusions regarding the efficacy and potency of these developing therapy strategies (summary of recent studies has been summarized in Table 1). In addition, it is important to note that most of these agents still need to be researched in clinical settings. At the time of writing this review article, there are three phase 2 clinical trials, NCT0296892 (everolimus, auranofin, vitamin D, CC-11050), NCT03160638 (azithromycin), and NCT03281226 (N-acetylcysteine), that are active, with two of them recruiting patients to demonstrate the efficacies of these various HDT therapy strategies. These data sets can potentially change current anti-TB protocols and induce a shift in direction for anti-TB therapies.

Table 1. Host-Directed Therapies against TB: Recent Clinical and Additional Studies.

Candidate	Description	Results	Remarks	References
		1. Clinical Studies		
Everolimus, Auranofin, Vitamin D, CC-11050	Combination therapy of HDT with DOTS drug regimen, followed with a modified DOTs protocol for 4 months with the intent to improve efficacy and outcomes of TB	Active, not enrolling.	Randomized, phase 2 clinical trial in South Africa	ClinicalTrials.gov Identifier: NCT02968927 [48,49]
Azithromycin	Immunomodulatory, adjunctive HDT therapy on top of the current DOTs regimen to reduce excessive inflammation, tissue degradation, and improve clinical outcomes of TB	Active, enrolling	Prospective, randomized, phase 2 pilot study in the Netherlands	ClinicalTrials.gov Identifier: NCT03160638
N-acetylcysteine	N-acetylcysteine in conjunction with rifampicin, isoniazid, pyrazinamide, ethambutol to provide anti-TB and antioxidative effects for patients with active HIV/TB infections	Active, enrolling	Randomized, phase 2 clinical trial in Brazil	ClinicalTrials.gov Identifier: NCT03281226
Vitamin D	Adjunctive vitamin D therapy in combination to standard antibiotic treatment for pulmonary tuberculosis to potentially enhance patient response	Vitamin D supplementation did not significantly reduce sputum conversion time among study population	Double-blind, randomized phase 3 clinical trial in the United Kingdom	ClinicalTrials.gov Identifier: NCT00419068 [50]
	Vitamin D supplementation to standard DOTs therapy with the hopes of quicker patient recovery times (demonstrated by sputum culture conversion)	Vitamin D supplementation did not significantly reduce sputum conversion time	Double-blind, randomized, placebo-controlled phase 3 clinical trial in South India	ClinicalTrials.gov Identifier: NCT00366470 [51]
	Effects of adjunctive vitamin D on host immunity with respect to TB and response to appropriate treatment	High-dose vitamin D3 corrected deficiency among patient, but did not improve TB clearance over the course of the trial	Double-blind, randomized, controlled phase 2 clinical trial in the United States	ClinicalTrials.gov Identifier: NCT00918086 [52]
	Determining if replacement of vitamin D in deficient patients with active TB affects clinical outcome	Vitamin D in high doses resulted in improvement in all TB patients, including those with vitamin D deficiencies.	Randomized, placebo-controlled clinical trial in Pakistan	ClinicalTrials.gov Identifier: NCT01130311 [53,54]
	Vitamin D and L-arginine supplementation in diagnosed TB patients in order to improve clinical outcomes and responses to pulmonary TB	With the doses administered, neither vitamin D nor L-arginine supplementation affected TB outcomes	Randomized, double-blind, placebo-controlled phase 3 clinical trial in Indonesia	ClinicalTrials.gov Identifier: NCT00677339 [55]
	Vitamin D3 and phenylbutyrate supplementation to standard short course DOTS therapy in order to improve recovery times and improve clinical outcomes in newly diagnosed TB patients	Beneficial effects towards patient recovery has been observed with phenylbutyrate, vitamin D3, or combination of phenylbutyrate and vitamin D3 supplementation with standard short-course therapy	Randomized, double-blind, placebo-controlled, 4-arm Phase 2 clinical trial in Bangladesh	ClinicalTrials.gov Identifier: NCT01580007 [28]

Table 1. *Cont.*

Candidate	Description	Results	Remarks	References
		2. Additional Studies		
Metformin	Multiple studies investigating the supplementation of metformin to existing standard anti-tuberculosis therapies, specifically in application to diabetic-TB patients	Metformin as an adjunctive therapy for diabetic TB patients needs to be understood further, even at the clinical level, due to inconsistent outcome reporting across studies	Retrospective cohort or case control studies. While there are reports of positive effects of metformin on active TB infections, there has been reports of no significant benefits in utilizing metformin as an adjunctive therapy.	[38,39,56–59]
Statins	Studies sought to understand the usage of cholesterol-lowering lipids (i.e., statins) and outcomes in regard to TB infections	Statins show beneficial effects as adjunctive therapy in TB infected *M. marinum* TB, and have been observed to shorten the culture negativity, reduce tissue pathology, and enhance bacterial killing along standard TB therapy. However, statins did not prevent TB progression in individuals who were newly diagnosed with type 2 diabetes mellitus.	Further studies will need to be conducted in order to understand the effects of statins and other potential cholesterol-lowering drugs on recovery times and outcome improvement in TB infections.	[42]

By modulating key host proteins, there remains the possibility of adverse unintended consequences. Nonetheless, novel approaches using HDTs can also provide with benefits in additional contexts, including immune responses to other organisms that may not pertain to TB infections. Additionally, it is just as important to identify new biomarkers to diagnose TB earlier with the hopes of preventing disease progression. Current TB diagnostic methods present with serious limitations in either precision or efficacy. This delays the diagnosis of TB significantly, further aiding in the complexity of the disease. With current advancements in biosensing technologies and point-of-care diagnostic tools, there is large potential to improve detection of TB. These approaches to managing and diagnosing TB, along with advancements in genomics and technology, can result in precise and effective novel therapies with the hopes of mitigating the global burden mycobacterial infections present. From a clinical perspective, these approaches are still in a relative infancy.

Author Contributions: Y.D., D.V., K.S. (Kush Shah) and K.S. (Kevin Shin) have equally contributed to drafting this review. V.V. conceived the frame work, provided guidance and assistance, and made edits to the draft.

Funding: This study was supported by the National Heart Blood Lung Institute at the National Institutes of Health (NIH) award 1R15HL143545-01A1.

Conflicts of Interest: The authors declare no conflict of interest.

References

1. Global Tuberculosis Report 2018. Available online: https://www.who.int/tb/publications/global_report/en/ (accessed on 26 June 2019).
2. Mittal, C.; Gupta, S. Noncompliance to DOTS: How it can be decreased. *Indian J. Community Med.* **2011**, *36*, 27–30. [CrossRef] [PubMed]
3. Sharma, V.; Verma, S.; Seranova, E.; Sarkar, S.; Kumar, D. Selective Autophagy and Xenophagy in Infection and Disease. *Front. Cell Dev. Biol.* **2018**, *6*, 147. [CrossRef] [PubMed]
4. Cerni, S.; Shafer, D.; To, K.; Venketaraman, V. Investigating the Role of Everolimus in mTOR Inhibition and Autophagy Promotion as a Potential Host-Directed Therapeutic Target in Mycobacterium tuberculosis Infection. *J. Clin. Med.* **2019**, *8*, 232. [CrossRef] [PubMed]

5. Nazio, F.; Strappazzon, F.; Antonioli, M.; Bielli, P.; Cianfanelli, V.; Bordi, M.; Gretzmeier, C.; Dengjel, J.; Piacentini, M.; Fimia, G.M.; et al. mTOR inhibits autophagy by controlling ULK1 ubiquitylation, self-association and function through AMBRA1 and TRAF6. *Nat. Cell Biol.* **2013**, *15*, 406–416. [CrossRef] [PubMed]
6. Lachmandas, E.; Beigier-Bompadre, M.; Cheng, S.; Kumar, V.; van Laarhoven, A.; Wang, X.; Ammerdorffer, A.; Boutens, L.; de Jong, D.; Kanneganti, T.; et al. Rewiring cellular metabolism via the AKT/mTOR pathway contributes to host defense against Mycobacterium tuberculosis in human and murine cells. *Eur. J. Immunol.* **2016**, *46*, 2574–2586. [CrossRef]
7. Wallis, R.S.; Hafner, R. Advancing host-directed therapy for tuberculosis. *Nat. Rev. Immunol.* **2015**, *15*, 255–263. [CrossRef] [PubMed]
8. Pagan, A.J.; Levitte, S.; Berg, R.D.; Hernandez, L.; Zimmerman, J.; Tobin, D.M.; Ramakrishnan, L. mTOR deficiency reveals an immunological trade-off in innate resistance to mycobacterial infection in vivo. *J. Immunol.* **2016**, *196* (Suppl. 1), 1.
9. Harrison, D.E.; Strong, R.; Sharp, Z.D.; Nelson, J.F.; Astle, C.M.; Flurkey, K.; Nadon, N.L.; Wilkinson, J.E.; Frenkel, K.; Carter, C.S.; et al. Rapamycin fed late in life extends lifespan in genetically heterogeneous mice. *Nature* **2009**, *460*, 392–395. [CrossRef]
10. Jagannath, C.; Lindsey, D.R.; Dhandayuthapani, S.; Xu, Y.; Hunter, R.L.; Eissa, N.T. Autophagy enhances the efficacy of BCG vaccine by increasing peptide presentation in mouse dendritic cells. *Nat. Med.* **2009**, *15*, 267–276. [CrossRef]
11. Andersson, A.-M.; Andersson, B.; Lorell, C.; Raffetseder, J.; Larsson, M.; Blomgran, R. Autophagy induction targeting mTORC1 enhances Mycobacterium tuberculosis replication in HIV co-infected human macrophages. *Sci. Rep.* **2016**, *6*, 28171. [CrossRef]
12. Mannick, J.B.; Del Giudice, G.; Lattanzi, M.; Valiante, N.M.; Praestgaard, J.; Huang, B.; Lonetto, M.A.; Maecker, H.T.; Kovarik, J.; Carson, S.; et al. mTOR inhibition improves immune function in the elderly. *Sci. Transl. Med.* **2014**, *6*, 268. [CrossRef] [PubMed]
13. Fijałkowska-Morawska, J.B.; Jagodzińska, M.; Nowicki, M. Pulmonary embolism and reactivation of tuberculosis during everolimus therapy in a kidney transplant recipient. *Ann. Transplant.* **2011**, *16*, 107–110. [CrossRef] [PubMed]
14. Singh, P.; Subbian, S. Harnessing the mTOR Pathway for Tuberculosis Treatment. *Front. Microbiol.* **2018**, *9*, 70. [CrossRef] [PubMed]
15. Gupta, A.; Pant, G.; Mitra, K.; Madan, J.; Chourasia, M.K.; Misra, A. Inhalable Particles Containing Rapamycin for Induction of Autophagy in Macrophages Infected with Mycobacterium tuberculosis. *Mol. Pharm.* **2014**, *11*, 1201–1207. [CrossRef] [PubMed]
16. Kahlenberg, J.M.; Kaplan, M.J. Little Peptide, Big Effects: The Role of LL-37 in Inflammation and Autoimmune Disease. *J. Immunol.* **2013**, *191*, 4895–4901. [CrossRef] [PubMed]
17. Shin, D.-M.; Jo, E.-K. Antimicrobial Peptides in Innate Immunity against Mycobacteria. *Immune Netw.* **2011**, *11*, 245–252. [CrossRef]
18. Rivas-Santiago, B.; Hernandez-Pando, R.; Carranza, C.; Juarez, E.; Contreras, J.L.; Aguilar-Leon, D.; Torres, M.; Sada, E. Expression of Cathelicidin LL-37 during Mycobacterium tuberculosis Infection in Human Alveolar Macrophages, Monocytes, Neutrophils, and Epithelial Cells. *Infect. Immun.* **2007**, *76*, 935–941. [CrossRef] [PubMed]
19. Stephan, A.; Batinica, M.; Steiger, J.; Hartmann, P.; Zaucke, F.; Bloch, W.; Fabri, M. LL37:DNA complexes provide antimicrobial activity against intracellular bacteria in human macrophages. *Immunology* **2016**, *148*, 420–432. [CrossRef]
20. Gupta, S.; Winglee, K.; Gallo, R.; Bishai, W.R. Bacterial subversion of cAMP signalling inhibits cathelicidin expression, which is required for innate resistance to Mycobacterium tuberculosis. *J. Pathol.* **2017**, *242*, 52–61. [CrossRef]
21. Napier, B.A.; Burd, E.M.; Satola, S.W.; Cagle, S.M.; Ray, S.M.; Mcgann, P.; Pohl, J.; Lesho, E.P.; Weiss, D.S. Clinical Use of Colistin Induces Cross-Resistance to Host Antimicrobials in Acinetobacter baumannii. *MBio* **2013**, *4*, e00021-13. [CrossRef]
22. Liu, P.T. Toll-Like Receptor Triggering of a Vitamin D-Mediated Human Antimicrobial Response. *Science* **2006**, *311*, 1770–1773. [CrossRef] [PubMed]

23. He, C.-S.; Yong, X.H.A.; Walsh, N.P.; Gleeson, M. Is there an optimal vitamin D status for immunity in athletes and military personnel? *Exerc. Immunol. Rev.* **2016**, *22*, 42–64. [PubMed]
24. Yuk, J.-M.; Shin, D.-M.; Lee, H.-M.; Yang, C.-S.; Jin, H.S.; Kim, K.-K.; Lee, Z.-W.; Lee, S.-H.; Kim, J.-M.; Jo, E.-K. Vitamin D3 Induces Autophagy in Human Monocytes/Macrophages via Cathelicidin. *Cell Host Microbe* **2009**, *6*, 231–243. [CrossRef] [PubMed]
25. Liu, P.T.; Stenger, S.; Tang, D.H.; Modlin, R.L. Cutting edge: Vitamin D-mediated human antimicrobial activity against Mycobacterium tuberculosis is dependent on the induction of cathelicidin. *J. Immunol.* **2007**, *179*, 2060–2063. [CrossRef] [PubMed]
26. Rook, G.A.; Steele, J.; Fraher, L.; Barker, S.; Karmali, R.; O'Riordan, J.; Stanford, J. Vitamin D3, gamma interferon, and control of proliferation of Mycobacterium tuberculosis by human monocytes. *Immunology* **1986**, *57*, 159–163. [PubMed]
27. Musarurwa, C.; Zijenah, L.S.; Mhandire, D.Z.; Bandason, T.; Mhandire, K.; Chipiti, M.M.; Munjoma, M.W.; Mujaji, W.B. Higher serum 25-hydroxyvitamin D concentrations are associated with active pulmonary tuberculosis in hospitalised HIV infected patients in a low income tropical setting: A cross sectional study. *BMC Pulm. Med.* **2018**, *18*, 67. [CrossRef] [PubMed]
28. Mily, A.; Rekha, R.S.; Kamal, S.M.M.; Arifuzzaman, A.S.M.; Rahim, Z.; Khan, L.; Haq, M.A.; Zaman, K.; Bergman, P.; Brighenti, S.; et al. Significant Effects of Oral Phenylbutyrate and Vitamin D3 Adjunctive Therapy in Pulmonary Tuberculosis: A Randomized Controlled Trial. *PLoS ONE* **2015**, *10*, e0138340. [CrossRef]
29. Sudfeld, C.R.; Mugusi, F.; Aboud, S.; Nagu, T.J.; Wang, M.; Fawzi, W.W. Efficacy of vitamin D3 supplementation in reducing incidence of pulmonary tuberculosis and mortality among HIV-infected Tanzanian adults initiating antiretroviral therapy: Study protocol for a randomized controlled trial. *Trials* **2017**, *18*, 66. [CrossRef]
30. Jarczak, J.; Kosciuczuk, E.M.; Lisowski, P.; Strzałkowska, N.; Jozwik, A.; Horbańczuk, J.O.; Krzyżewski, J.; Zwierzchowski, L.; Bagnicka, E. Defensins: Natural component of human innate immunity. *Hum. Immunol.* **2013**, *74*, 1069–1079. [CrossRef]
31. Rivas-Santiago, B.; Schwander, S.K.; Sarabia, C.; Diamond, G.; Klein-Patel, M.E.; Hernández-Pando, R.; Ellner, J.J.; Sada, E. Human β-Defensin 2 Is Expressed and Associated with Mycobacterium tuberculosis during Infection of Human Alveolar Epithelial Cells. *Infect. Immun.* **2005**, *73*, 4505–4511. [CrossRef]
32. Miyakawa, Y.; Ratnakar, P.; Rao, A.G.; Costello, M.L.; Mathieu-Costello, O.; Lehrer, R.I.; Catanzaro, A. In vitro activity of the antimicrobial peptides human and rabbit defensins and porcine leukocyte protegrin against Mycobacterium tuberculosis. *Infect. Immun.* **1996**, *64*, 926–932. [PubMed]
33. Dong, H.; Lv, Y.; Zhao, D.; Barrow, P.; Zhou, X. Defensins: The Case for Their Use against Mycobacterial Infections. *J. Immunol. Res.* **2016**, *2016*, 1–9. [CrossRef] [PubMed]
34. Sharma, S.; Verma, I.; Khuller, G.K. Therapeutic Potential of Human Neutrophil Peptide 1 against Experimental Tuberculosis. *Antimicrob. Agents Chemother.* **2001**, *45*, 639–640. [CrossRef] [PubMed]
35. Rivas-Santiago, C.E.; Rivas-Santiago, B.; Leon, D.A.; Castañeda-Delgado, J.; Pando, R.H. Induction of β-defensins by l-isoleucine as novel immunotherapy in experimental murine tuberculosis. *Clin. Exp. Immunol.* **2011**, *164*, 80–89. [CrossRef] [PubMed]
36. Lachmandas, E.; Eckold, C.; Böhme, J.; Koeken, V.A.C.M.; Marzuki, M.B.; Blok, B.; Arts, R.J.W.; Chen, J.; Teng, K.W.W.; Ratter, J.; et al. Metformin Alters Human Host Responses to Mycobacterium tuberculosis in Healthy Subjects. *J. Infect. Dis.* **2019**, *220*, 139–150. [CrossRef]
37. Singhal, A.; Jie, L.; Kumar, P.; Hong, G.S.; Leow, M.K.-S.; Paleja, B.; Tsenova, L.; Kurepina, N.; Chen, J.; Zolezzi, F.; et al. Metformin as adjunct antituberculosis therapy. *Sci. Transl. Med.* **2014**, *6*, 263. [CrossRef] [PubMed]
38. Marupuru, S.; Senapati, P.; Pathadka, S.; Miraj, S.S.; Unnikrishnan, M.K.; Manu, M.K. Protective effect of metformin against tuberculosis infections in diabetic patients: An observational study of south Indian tertiary healthcare facility. *Braz. J. Infect. Dis.* **2017**, *21*, 312–316. [CrossRef]
39. Lee, M.-C.; Chiang, C.-Y.; Lee, C.-H.; Ho, C.-M.; Chang, C.-H.; Wang, J.-Y.; Chen, S.-M. Metformin use is associated with a low risk of tuberculosis among newly diagnosed diabetes mellitus patients with normal renal function: A nationwide cohort study with validated diagnostic criteria. *PLoS ONE* **2018**, *13*, e0205807. [CrossRef]

40. Ma, S.; Ma, C.C.-H. Recent development in pleiotropic effects of statins on cardiovascular disease through regulation of transforming growth factor-beta superfamily. *Cytokine Growth Factor Rev.* **2011**, *22*, 167–175. [CrossRef]
41. Parihar, S.P.; Guler, R.; Khutlang, R.; Lang, D.M.; Hurdayal, R.; Mhlanga, M.M.; Suzuki, H.; Marais, A.D.; Brombacher, F. Statin Therapy Reduces the Mycobacterium tuberculosis Burden in Human Macrophages and in Mice by Enhancing Autophagy and Phagosome Maturation. *J. Infect. Dis.* **2013**, *209*, 754–763. [CrossRef]
42. Skerry, C.; Pinn, M.L.; Bruiners, N.; Pine, R.; Gennaro, M.L.; Karakousis, P.C. Simvastatin increases the in vivo activity of the first-line tuberculosis regimen. *J. Antimicrob. Chemother.* **2014**, *69*, 2453–2457. [CrossRef]
43. Kang, Y.A.; Choi, N.-K.; Seong, J.-M.; Heo, E.Y.; Koo, B.K.; Hwang, S.-S.; Park, B.-J.; Yim, J.-J.; Lee, C.-H. The effects of statin use on the development of tuberculosis among patients with diabetes mellitus. *Int. J. Tuberc. Lung Dis.* **2014**, *18*, 717–724. [CrossRef]
44. Anisimova, Y.; Gelperina, S.; Peloquin, C.; Heifets, L. Nanoparticles as Antituberculosis Drugs Carriers: Effect on Activity Against Mycobacterium tuberculosis in Human Monocyte-Derived Macrophages. *J. Nanopart. Res.* **2000**, *2*, 165–171. [CrossRef]
45. Vibe, C.B.; Fenaroli, F.; Pires, D.; Wilson, S.R.; Bogoeva, V.; Kalluru, R.; Speth, M.; Anes, E.; Griffiths, G.; Hildahl, J. Thioridazine in PLGA nanoparticles reduces toxicity and improves rifampicin therapy against mycobacterial infection in zebrafish. *Nanotoxicology* **2015**, *10*, 680–688. [CrossRef]
46. Mathur, M.L. Role of vitamin A supplementation in the treatment of tuberculosis. *Natl. Med. J. India* **2007**, *20*, 16–21.
47. Polena, H.; Boudou, F.; Tilleul, S.; Dubois-Colas, N.; Lecointe, C.; Rakotosamimanana, N.; Pelizzola, M.; Andriamandimby, S.F.; Raharimanga, V.; Charles, P.; et al. Mycobacterium tuberculosis exploits the formation of new blood vessels for its dissemination. *Sci. Rep.* **2016**, *6*, 33162. [CrossRef]
48. Harbut, M.B.; Vilchèze, C.; Luo, X.; Hensler, M.E.; Guo, H.; Yang, B.; Chatterjee, A.K.; Nizet, V.; Jacobs, W.R.; Schultz, P.G.; et al. Auranofin Exerts Broad-Spectrum Bactericidal Activities by Targeting Thiol-Redox Homeostasis. *Proc. Natl. Acad. Sci. USA* **2015**, *112*, 4453–4458. [CrossRef]
49. Lin, K.; Obrien, K.M.; Trujillo, C.; Wang, R.; Wallach, J.B.; Schnappinger, D.; Ehrt, S. Mycobacterium Tuberculosis Thioredoxin Reductase Is Essential for Thiol Redox Homeostasis but Plays a Minor Role in Antioxidant Defense. *PLoS Pathog.* **2016**, *12*, e1005675. [CrossRef]
50. Martineau, A.R.; Timms, P.M.; Bothamley, G.H.; Hanifa, Y.; Islam, K.; Claxton, A.P.; Packe, G.E.; Moore-Gillon, J.C.; Darmalingam, M.; Davidson, R.N.; et al. High-Dose Vitamin D3 during Intensive-Phase Antimicrobial Treatment of Pulmonary Tuberculosis: A Double-Blind Randomised Controlled Trial. *Lancet* **2011**, *377*, 242–250. [CrossRef]
51. Daley, P.; Jagannathan, V.; John, K.R.; Sarojini, J.; Latha, A.; Vieth, R.; Suzana, S.; Jeyaseelan, L.; Christopher, D.J.; Smieja, M.; et al. Adjunctive Vitamin D for Treatment of Active Tuberculosis in India: A Randomised, Double-Blind, Placebo-Controlled Trial. *Lancet Infect. Dis.* **2015**, *15*, 528–534. [CrossRef]
52. Tukvadze, N.; Sanikidze, E.; Kipiani, M.; Hebbar, G.; Easley, K.A.; Shenvi, N.; Kempker, R.R.; Frediani, J.K.; Mirtskhulava, V.; Alvarez, J.A.; et al. High-Dose Vitamin D3 in Adults with Pulmonary Tuberculosis: A Double-Blind Randomized Controlled Trial. *Am. J. Clin. Nutr.* **2015**, *102*, 1059–1069. [CrossRef]
53. Hasan, Z.; Salahuddin, N.; Rao, N.; Aqeel, M.; Mahmood, F.; Ali, F.; Ashraf, M.; Rahman, F.; Mahmood, S.; Islam, M.; et al. Change in serum CXCL10 levels during anti-tuberculosis treatment depends on vitamin D status [Short Communication]. *Int. J. Tuberc. Lung Dis.* **2014**, *18*, 466–469. [CrossRef]
54. Salahuddin, N.; Ali, F.; Hasan, Z.; Rao, N.; Aqeel, M.; Mahmood, F. Vitamin D Accelerates Clinical Recovery from Tuberculosis: Results of the SUCCINCT Study [Supplementary Cholecalciferol in Recovery from Tuberculosis]. A Randomized, Placebo-Controlled, Clinical Trial of Vitamin D Supplementation in Patients with Pulmonary Tuberculosis'. *BMC Infect. Dis.* **2013**, *13*, 22.
55. Ralph, A.P.; Waramori, G.; Pontororing, G.J.; Kenangalem, E.; Wiguna, A.; Tjitra, E.; Sandjaja; Lolong, D.B.; Yeo, T.W.; Chatfield, M.D.; et al. L-Arginine and Vitamin D Adjunctive Therapies in Pulmonary Tuberculosis: A Randomised, Double-Blind, Placebo-Controlled Trial. *PLoS ONE* **2013**, *8*, e70032. [CrossRef]
56. Ma, Y.; Pang, Y.; Shu, W.; Liu, Y.-H.; Ge, Q.-P.; Du, J.; Li, L.; Gao, W.-W. Metformin Reduces the Relapse Rate of Tuberculosis Patients with Diabetes Mellitus: Experiences from 3-Year Follow-Up. *Eur. J. Clin. Microbiol. Infect. Dis.* **2018**, *37*, 1259–1263. [CrossRef]

57. Novita, B.D.; Soediono, E.I.; Nugraha, J. Metformin associated inflammation levels regulation in type 2 diabetes mellitus-tuberculosis coinfection patients—A case report. *Indian J. Tuberc.* **2018**, *65*, 345–349. [CrossRef]
58. Novita, B.D.; Ali, M.; Pranoto, A.; Soediono, E.I.; Mertaniasih, N.M. Metformin induced autophagy in diabetes mellitus—Tuberculosis co-infection patients: A case study. *Indian J. Tuberc.* **2019**, *66*, 64–69. [CrossRef]
59. Degner, N.R.; Wang, J.-Y.; Golub, J.E.; Karakousis, P.C. Metformin Use Reverses the Increased Mortality Associated With Diabetes Mellitus During Tuberculosis Treatment. *Clin. Infect. Dis.* **2017**, *66*, 198–205. [CrossRef]

Journal of *Clinical Medicine*

MDPI

Review

Investigating the Role of Everolimus in mTOR Inhibition and Autophagy Promotion as a Potential Host-Directed Therapeutic Target in *Mycobacterium tuberculosis* Infection

Stephen Cerni [1,†], Dylan Shafer [1,†], Kimberly To [2] and Vishwanath Venketaraman [1,2,*]

1 Department of Basic Medical Sciences, College of Osteopathic Medicine of the Pacific, Western University of Health Sciences, Pomona, CA 91766-1854, USA; stephen.cerni@westernu.edu (S.C.) dylan.shafer@westernu.edu (D.S.)

2 Graduate College of Biomedical Sciences, Western University of Health Sciences, Pomona, CA 91766-1854, USA; kimberly.to@westernu.edu

* Correspondence: vvenketaraman@westernu.edu; Tel.: +1-909-706-3736

† These authors contributed equally to this work.

Received: 13 January 2019; Accepted: 8 February 2019; Published: 11 February 2019

check for updates

Abstract: Tuberculosis (TB) is a serious infectious disease caused by the pathogen *Mycobacterium tuberculosis* (*Mtb*). The current therapy consists of a combination of antibiotics over the course of four months. Current treatment protocols run into problems due to the growing antibiotic resistance of *Mtb* and poor compliance to the multi-drug-resistant TB treatment protocol. New treatments are being investigated that target host intracellular processes that could be effective in fighting *Mtb* infections. Autophagy is an intracellular process that is involved in eliminating cellular debris, as well as intracellular pathogens. Mammalian target of rapamycin (mTOR) is an enzyme involved in inhibiting this pathway. Modulation of mTOR and the autophagy cellular machinery are being investigated as potential therapeutic targets for novel *Mtb* treatments. In this review, we discuss the background of *Mtb* pathogenesis, including its interaction with the innate and adaptive immune systems, the mTOR and autophagy pathways, the interaction of *Mtb* with these pathways, and finally, the drug everolimus, which targets these pathways and is a potential novel therapy for TB treatment.

Keywords: *Mycobacterium tuberculosis*; host-directed therapies; immune responses

1. Introduction

Tuberculosis (TB) is an ancient infectious disease caused by *Mycobacterium tuberculosis* (*Mtb*) that still plagues the modern world. *Mtb* has survived over 70,000 years, and today actively infects around 10 million people annually and lies latent in 1.7 billion people worldwide (23% of the global population) [1]. Claiming over a million lives a year, TB is the leading cause of death by an infectious agent over human immunodeficiency virus/acquired immunodeficiency syndrome (HIV/AIDS). Additionally, immunocompromised individuals, such as those with HIV and type 2 diabetes (T2DM) are at a greater risk of developing active TB. The common treatment for drug-sensitive pulmonary TB by the World Health Organization (WHO) is the Directly Observed Treatment, Short Course (DOTS). DOTS is comprised of an antibiotic regimen of isoniazid (INH), rifampicin (RIF), pyrazinamide (PZA), and ethambutol (ETH) in the initial phase for 2 months, followed by INH and RIF in the continuation phase for 4 months. DOTS therapy is currently the best curative treatment for TB, but the long duration and potential adverse side-effects cause a high non-compliance/drop-out rate. Patient non-compliance increases the risks for development of drug-resistant TB and contributes to TB's status as one of the top ten causes of death globally [2]. TB's continuous threat to public health warrants investigation into more effective treatments.

A relatively new modality of TB treatment is called Host Directed Therapy (HDT). HDT aims to augment the endogenous host immune system to battle TB infection, through the use of pharmacology [3]. One target of interest for HDT in TB treatment is autophagy. Autophagy is an intracellular homeostatic process that degrades damaged cellular components and organelles during times of cellular stress via lysosomal degradation [4]. This process is also part of innate immunity and is involved in eliminating intracellular pathogens. Autophagy is also involved in adaptive immunity and might facilitate antigen presentation, which eventually leads to granuloma formation. *Mtb* is able to hinder the host cells' ability to complete autophagy, through the modulation of mammalian target of rapamycin (mTOR). Everolimus, a potential HDT, might be able to modulate this effect on mTOR and could be a novel treatment for *Mtb*. Here, we have investigated the role of mTOR in the intracellular autophagy of *Mtb* and its implication as a target for future treatment.

2. Autophagy Overview

Autophagy is a homeostatic cellular process that involves removing protein aggregates and damaged organelles via lysosomal degradation. This process is crucial for cells to survive under stressful conditions and involves eliminating unnecessary or damaged elements from the cell [5]. It is also a key process for removing invading pathogens, making it a potential target for directed therapies [4,6]. Autophagy has many different subtypes based on the target of degradation and can be selective (for a particular organelle or pathogen) or non-selective (also referred to as macro autophagy or bulk autophagy). For the purposes of this review, we will focus on xenophagy, which is a type of selective autophagy that specifically targets intracellular pathogens [5]. We will review the general process of autophagy as well as specific autophagic processes as they pertain to *Mtb*, including interactions with the innate and adaptive immune systems. An investigation of the relationship between autophagy and *Mtb* is critical in understanding the potential targets of HDT.

Autophagy begins with the formation of an autophagosome, which is a double-membrane-bound vesicle that contains cytoplasmic material [4]. These autophagosomes are non-degradative until they come in contact with lysosomes, forming an autolysosome, which enables them to degrade their contents [4,7]. The induction of autophagy is complex but involves three main components, the phosphoinositide 3-kinase complex 3 (PI3KC3), Unc-51-like Kinase 1 complex (ULK1), and the autophagy-related protein (ATG) complex [6]. The process of autophagy is inhibited by the mTOR complex [6], which is a focus for potential *Mtb* therapeutics. The specific mechanisms of this interaction will be discussed later in this review. Autophagy is not a single pathway and has many effects, both, with the innate immune system and the adaptive immune system. In this review, three autophagy pathways will be discussed: direct pathogen degradation (also referred to as xenophagy), interaction with the innate immune system, and interactions with the adaptive immune system [7].

Xenophagy is a specific type of autophagy that describes the process of delivering intracellular pathogens to lysosomes via autophagic mechanisms [4]. The precise mechanisms of xenophagy are not well understood; however, there are several proposed hypotheses [7]. An overview of xenophagy can be seen in Figure 1. There are three general steps in the autophagy pathway: initiation, elongation of the autophagosome, and maturation of the autophagolysosome and degradation of its contents.

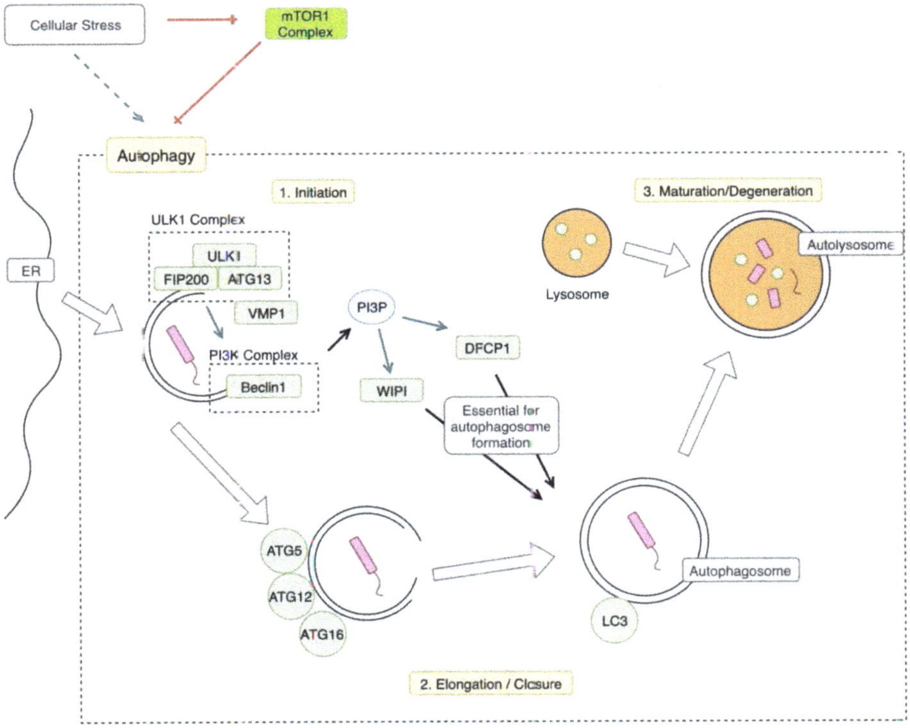

Figure 1. Xenophagy Pathway Overview. Cellular stress including starvation or hypoxia can trigger the autophagy pathway by relieving inhibition by mammalian target of rapamycin 1 (mTOR1). In the case of xenophagy, autophagic mechanisms are triggered by an intracellular pathogen. The pathway begins with phosphorylation of the Unc-51-like Kinase 1 (ULK1) complex, which activates the phosphoinositide 3-kinase complex 3 (PI3KC3) complex. This begins the double-membraned autophagosome formation, which is derived from the endoplasmic reticulum (ER). The next step is elongation and closure of the autophagosome. An autophagy-related protein (ATG) complex comprised of ATG5, ATG12, and ATG16 is involved in this next step. Microtubule-associated protein 1A/1B-light chain 3 (LC3) is also conjugated to the membrane, at this step. Phosphatidylinositol 3 phosphate (PI3P) produced by the PI3KC3 complex is necessary for autophagosome closure, as well. The final step is fusion with a lysosome forming an autolysosome, degrading its contents. VMP1: vacuole membrane protein 1, FIP200: focal adhesion kinase family interacting protein of 200kd, WIPI: WD repeat domain phosphoinositide-interacting protein 1, DFCP1: double FYVE domain-containing protein 1.

Before the autophagy mechanisms begin, the intracellular pathogen is tagged with ubiquitin. The process of tagging intracellular bacteria is similar to that of tagging endogenous proteins or organelles for destruction via ubiquitination [6]. Autophagy receptors play an important role in target recognition and delivery to the autophagosome [8]. These specific mechanisms will be discussed in the innate immune system section. The next step is the initiation of autophagosome formation. Generally, autophagy begins with inhibition of mTOR, which results in translocation of the mTOR to the endoplasmic reticulum (ER) and the subsequent phosphorylation of the ULK1 complex, inducing autophagy [9]. This leads to the recruitment of PI3KC3, which produces phosphatidylinositol 3 phosphate (PI3P); this is essential for the autophagosome formation via the double FYVE-containing protein 1 (DFCP1) and the beta transducin (WD)-repeat domain phosphoinositide-interacting (WIPI) proteins [8,10]. Vacuole Membrane Protein 1(VMP1) also appears to play a role in the autophagosome formation, via interaction with beclin 1 and the ULK1 complex [4,8]. The last step of the autophagosome formation requires an ATG12–ATG5 complex and a Microtubule-associated protein 1A/1B-light chain 3 (LC3). The final step of autophagy is the fusion of the autophagosome with the lysosomal compartment [4]. This leads to the formation of the autophagolysosome, which ultimately degrades the pathogen.

In addition to pathogen degradation via xenophagy, autophagy may modulate the innate and adaptive immune systems [4]. With respect to the innate immune system, autophagy can both enhance interferon production and prevent excessive innate immune system activity from being detrimental [4]. With respect to adaptive immunity, autophagy is reportedly involved with both the delivery of exogenous and endogenous antigens to major histocompatibility complex (MHC) class II antigen-presenting molecules and the presentation of viral antigens by MHC class I antigen presenting molecules [4].

3. Autophagy and TB in the Innate Immune System

There appear to be two autophagy pathways involved in an *Mtb* infection, which ultimately lead to the formation of an autolysosome. These pathways are illustrated in Figure 2. The first classical pathway is xenophagy, which is triggered by a mycobacterial infection via activation of the bacterial early secretory antigenic target (ESAT)-6 secretion system (ESX-1), leading to a disruption of the phagosome membrane [11,12]. The bacterial DNA is next recognized via the stimulator of interferon genes (STING)-dependent pathway which recognizes DNA on the surface of the bacteria [11]. This STING pathway then leads to the direct ubiquitination of the bacteria, which involves activity by the parkin ligase, the Smurf1 ligase, and the TRIM3 ligase [13–16]. The parkin ligase specifically mediates the linkage of K63 ubiquitin chains [14] and the Smurf1 ligase mediates the linkage of K48 ubiquitin chains [15]. Next, the ubiquitin LC3-binding autophagy adaptors p62 and NPD52 bind ubiquitin, leading to the recruitment of autophagic components, which create a phagophore around the bacteria [11].

An offshoot of this classical pathway involves the murine immunity-related p47 guanosine triphosphatase family M protein 1 (IRGM1). IRGM1 is thought to contribute to this pathway through stabilization of autophagy factors and adenosine monophosphate (AMP)-activated protein kinase (AMPK) [17]. IRGM1 is induced by similar factors that induce general autophagy, including starvation and IFN-γ [17].

The second pathway involved is called the LC3 associated phagocytosis (LAP) pathway [13]. This pathway differs from the classical autophagy pathway in that it does not involve a double membrane autophagosome, but rather a single membrane LAPsome [18]. The pathway begins with Toll-like receptor (TLR) activation by *Mtb*, which leads to phagocytosis of the pathogen [19]. The phagocytosed pathogen is now in a single-membrane LAPsome, which has PI3P attached to it (produced by PI3KC3) and includes key LAP proteins, such as Beclin 1 and Rubicon [13]. This leads to the generation of reactive oxygen species (ROS) via the stabilization of the nicotinamide adenine dinucleotide phosphate (NADPH) oxidase-2 (NOX2) complex [20]. The combination of PI3P, ROS, and several other ATGs lead to LC3 being conjugated to the single phagosome membrane [20,21]. This eventually leads to fusion with cellular lysosomes and the destruction of the pathogen [20].

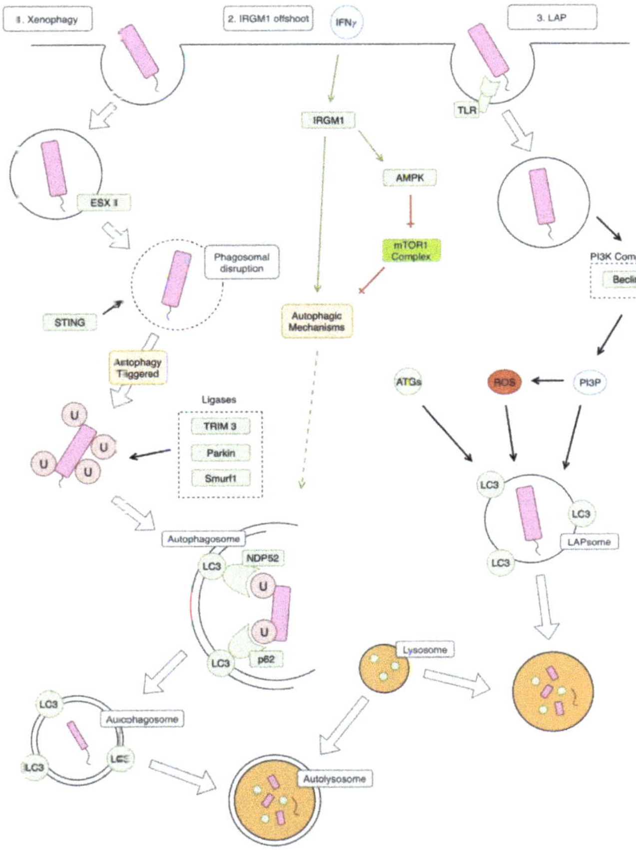

Figure 2. Autophagy and *Mycobacterium tuberculosis* (*Mtb*). The first pathway illustrated in the classic xenophagy pathway. The phagosome is disrupted by bacterial early secretory antigenic target (ESAT)-6 secretion system (ESX1). This allows for recognition of bacterial DNA by the stimulator of interferon genes (STING) pathway, which triggers autophagy. Several ligases are involved in the ubiquitination of the bacteria, including Tripartite motif-containing protein 3 (TRIM3), Parkin, and Smurf1. This ubiquitination facilitates recognition of the bacteria by autophagic mechanisms, leading to the formation of a double-membraned autophagosome. This autophagosome fuses with a lysosome leading to the formation of an autolysosome and the degradation of bacteria. A second pathway, and an offshoot of the first xenophagy pathway, involves immunity-related p47 guanosine triphosphatase family M protein 1 (IRGM1). IRGM1 is triggered by interferon-gamma (IFN-γ) and leads to the activation of autophagic mechanisms. The specifics of this pathway are unclear, but it might involve activation of adenosine monophosphate (AMP)-activated protein kinase (AMPK), which relieves the autophagy pathway of its inhibition by mTOR1. The third and final pathway is also referred to as the 1A/1B-light chain 3 (LC3) associated phagocytosis (LAP) pathway. Bacteria is recognized by a toll-like receptor (TLR) and phagocytosed. This process triggers the phosphoinositide 3-kinase (PI3K) complex, leading to the production of PI3P. PI3P leads to the production of reactive oxygen species (ROS), via stabilization of the nicotinamide adenine dinucleotide phosphate hydrogen (NADPH) oxidase-2 (NOX) complex. Phosphatidylinositol 3 phosphate (PI3P), ROS, and other autophagy related proteins (ATGs) lead to the formation of a LAPsome, which is a single-membraned compartment. Microtubule-associated protein 1A/1B-light chain 3 (LC3) is conjugated to the LAPsome, which triggers fusion with a lysosome, leading to degradation of the bacteria. NDP52: nuclear domain 10, protein 2.

The LAP pathway of autophagy can prevent *Mtb* from inhibiting the maturation of lysosomes. *Mtb* is phagocytosed by an actin-mediated membrane that engulfs the bacterium into a phagosome. Ideally, Rab GTPases activate within the phagosomal membrane to recruit vacuolar ATPases that acidify the phagosomal contents. Then, the phagosome proceeds to fuse with the lysosome which further acidifies and degrades bacteria via enzymatic processes. However, *Mtb* alters the phagosome trafficking pathway through numerous methods, preventing its own elimination by this process [22,23]. *Mtb* uses phosphatidylinositol mannoside (PIM) to stimulate Rab14, promoting phagosome fusion with an early endosome, resulting in the prevention of phagosomal maturation and acidification [24]. Other studies have found *Mtb* uses Lipoarabinomannan Mannosylated (ManLAM) to interfere with the calmodulin complex formation with PI3KC3 and the production of PI3P, which is responsible for the recruitment of vacuolar GTPases to the phagosome, preventing its maturation and fusion with the lysosome [25]. *Mtb* also prevents phagosomal maturation by preventing the phagosome from acquiring Rab5 due to the presence of tryptophan aspartate coat protein (TACO). Through these processes *Mtb* resides in the phagosome at a pH of 6.2 rather than the normal physiologic levels, which can reach a pH of <5.0 [26]. However, when autophagy is induced stronger PI3KC3 and PI3P activation occurs and *Mtb's* inhibition of lysosomal maturation is overcome [27].

4. Autophagy and TB in the Adaptive Immune System

An adaptive cellular immune response is needed to effectively control an *Mtb* infection. After the initial innate immune responses, adaptive immunity develops to control the dividing bacteria. Development of adaptive immune responses occur between 3 to 8 weeks after the initial exposure to *Mtb*. Impaired adaptive immunity often results in clinical TB. Effective host immune responses against *Mtb* infection are dependent on the optimal interactions between the appropriate T cell subsets and infected macrophages.

4.1. Type 1 T helper (TH1)

Immunity to *Mtb* infection is associated with the emergence of protective CD4+ T cells that secrete cytokines, resulting in the activation of macrophages and recruitment of monocytes for granuloma formation. Studies in human and animals have demonstrated that acquired immunity to *Mtb* involved multiple T cell subsets with a dominant role of CD4+ T helper cells and aid from the CD8+ T cells [28]. Type 1 T helper (Th1) cells produce interferon-gamma (IFN-γ), interleukin, (IL)-2, and tumor necrosis factor (TNF)-beta, which activate macrophages and are responsible for cell-mediated immunity and phagocyte-dependent protective responses. These CD4+ Th1 cells may recognize mycobacterial fragments by the presence of MHC II class molecules on the antigen-presenting cells, such as macrophages, although this concept is debated in recent literature as *Mtb* may have some mechanisms for inhibiting MHC II presentation [29]. Mice with deleted genes for CD4+ or MHC class II molecules are significantly susceptible to *Mtb* infection, strongly establishing the central protective role of the CD4+ T cells [30,31]. Additionally, loss of the CD4+ cell number and function, during the advanced stages of HIV infection, results in progressive primary infection, reactivation of endogenous *Mtb*, and increased susceptibility to re-infection [32].

Following phagocytosis of *Mtb* by macrophages and dendritic cells, IL-12 secretion is induced, driving the development of a Th1 response with the production of IFN-γ. IFN-γ is involved in the recruitment of T-cells, in the induction of expression of the MHC class II molecules, in the augmentation of antigen presenting cells (APCs), and in the control of *Mtb* growth [33]. Additionally, IFN-γ promotes cellular proliferation, cell adhesion, apoptosis, and autophagy [34]. IL-12 is crucial for generation of a protective immunity, with its main function being the induction of expression of IFN-γ and activation of antigen-specific lymphocytes. Mice with deleted IL-12p40 gene were more susceptible to infection, had increased bacterial burden, and decreased survival time compared to control mice [35]. IFN-γ has been established as the principal mediator for a protective immune response to *Mtb* infection. IFN-γ

knockout (GKO) mice formed defective granulomas and failed to produce nitrogen intermediates [36]. CD8+ cells also secrete IFN-γ but to a lesser extent than that of the CD4+ T cells [34].

4.2. Granuloma Formation

Granuloma formation is the hallmark immunopathology of TB, providing a microenvironment for the T cell activation of infected macrophages, to the inhibition of bacterial growth, and localization of the inflammatory and immune responses to the site of infection. Granuloma formation is largely dependent on T cell-mediated immune responses and macrophage-derived cytokines, such as IFN-γ and members of the TNF superfamily [37]. Immediate and sustained secretion of chemokines is essential for the recruitment, migration, and aggregation of monocytes and lymphocytes to form granulomas at the sites of *Mtb* infection [38]. Granulomas are composed of various immune cells, including macrophages, dendritic cells, T cell, fibroblasts, epithelioid histiocytes, giant cells, and natural killer cells [37]. The granuloma provides a physical barrier which encapsulates and prevents bacteria from spreading. This local environment allows these immune cells to interact and to effectively kill *Mtb*, which is achieved by macrophage activation and creating an oxygen and nutrient-deprived environment [39]. Critical to granuloma formation is tumor necrosis factor-alpha (TNF-α). Mice deficient in TNF-α or the 55 kDa TNF receptor, died a rapid death, and with a sustainably higher bacterial burden, compared to control mice [40]. Another study showed structural deficiencies in granulation formation in the TNF-α gene-targeted mice. Therefore, TNF-α has a central role in anti-TB immunity, through generation of structurally effective granulomatous response [41].

The failure of the macrophages to acquire mycobactericidal function is likely is due to the host's inability to generate a sufficient Th1 cell-mediated response. Individuals with compromised cell-mediated immunity, such as HIV-positive patients and diabetic patients, are highly susceptible to *Mtb* infection. A mechanism of immunosuppression is attributed to decreased levels of GSH, which has been shown in HIV and in individuals with T2DM [42,43]. Supplementation with GSH can help restore cytokine balance and enhanced granulomatous response [42–44]. GSH is an essential component of intracellular antioxidant systems and functions in the protection of cells against oxidative stress and in maintaining redox homeostasis [45]. GSH could be a potential adjunct therapy to antibiotics and new host-directed therapies in helping relieve oxidative stress in cells.

4.3. Autophagy and Adaptive Immunity

Autophagy contributes to the crosstalk between the innate and adaptive immune response in *Mtb* infection by enhancing antigen presentation. Autophagy is reportedly involved in both the delivery of exogenous and endogenous antigens to MHC class II antigen presenting molecules, along with the presentation of viral antigens by MHC class I antigen presenting molecules [46]. Various authors have shown how autophagy can contribute to the MHC class II presentation, which is particularly important in *Mtb* defense [47,48]. The antigenic contents of the autophagosomes are degraded when they fuse with lysosomes. Then, within the multi-vesicular MHC-II loading compartments (MIICs), the antigenic peptides are fashioned into MHC-II binding groves by the HLA-DM. Autophagy increases the MIIC turnover and strongly improves the MHC class II presentation to CD4+ T cells [7]. A study conducted by Jagannath C. et al., 2009 showed autophagy augmented the efficacy of the BCG vaccine in mice, by improving antigen presentation by antigen presenting cells [49].

In addition to antigen presentation, the adaptive immune system and autophagy maintain a synergistic relationship through the production of cytokines in the defense of *Mtb*. IFN-γ produced by Th1 cells induces autophagy [28,29]. In turn, autophagy has been shown to increase the production of TNF-α, IL-6, and IL-8 [7]. In mice, IFN-γ produced by Th1 cells promotes the expression of a p47 resistance GTPase, called IFN-γ-inducible protein (LRG-47) [30,31]. It is suspected that LRG-47 prompts the creation of autophagolysosomes in the defense of *Mtb* [30]. Multiple experiments have shown that inhibition of certain ATGs, like Beclin-1, and treatment with autophagy inhibitor 3-MA in both murine and human models showed reduction of TNF-α, IL-6, and IL-8 [32]. Each of these cytokines play an important role in the inflammatory response against *Mtb*: IL-8 helps recruit neutrophils, IL-6 stimulates production of acute phase reaction, and TNF-α is essential in the production of granuloma formation. These findings point towards autophagy as a potent regulator of host defense against *Mtb*.

5. mTOR

The mammalian target of rapamycin (mTOR) is a regulator of many cellular processes involved in growth and differentiation. It is involved in many anabolic pathways and blocks catabolic processes, such as autophagy [50]. mTOR is active when nutrients are readily available to the cell, and is inactivated during times of starvation, leading to the induction of autophagy, which helps the cell survive under these unfavorable conditions [51]. The pathway discussed in this review is also referred to as the Protein kinase B (AKT)/mTOR pathway; illustrated in Figure 3.

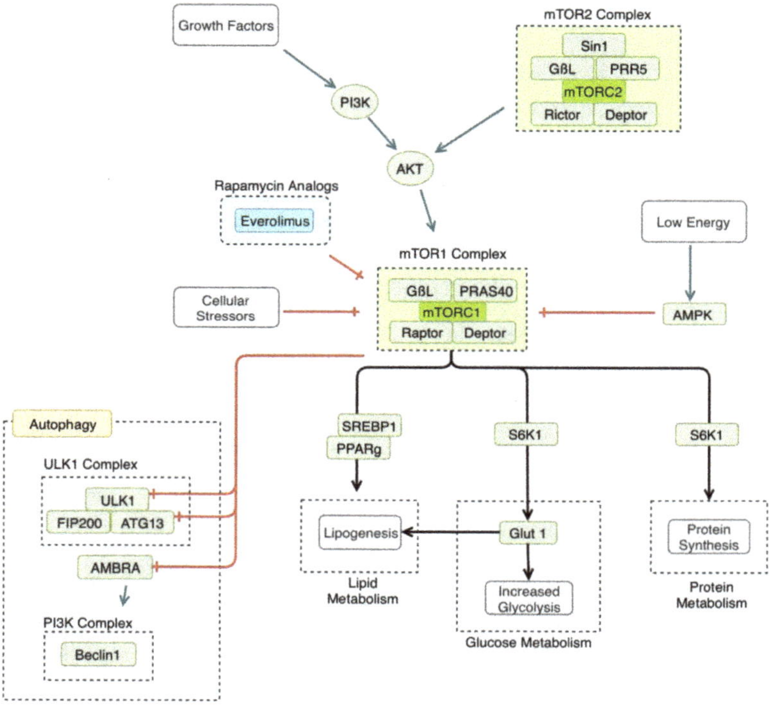

Figure 3. Mammalian target of rapamycin (mTOR) pathway. mTOR1 is triggered by certain growth factors and is generally anabolic, making it an important enzyme in both cancer and processes related to autophagy. Cellular stressors, such as hypoxia or cellular starvation, lead to the inhibition of mTOR, thus, activating the autophagy pathway. mTOR1 inhibits specific enzyme subunits in the autophagy pathway, including Unc-51-like Kinase 1 complex (ULK1), autophagy related protein 13 (ATG13), and activating molecule in Beclin 1-regulated autophagy protein 1 (AMBRA). mTOR1 activation has several metabolic downstream effects, including increased lipogenesis, increased glycolysis, and increased protein synthesis. mTOR is inhibited by compounds in the rapamycin pathway, including everolimus. PI3K: phosphoinositide 3-kinase, AKT: Protein kinase B, PRAS 40: proline-rich AKT substrate of 40kd, PRR5: proline-rich protein 5, AMPK: adenosine monophosphate (AMP)-activated protein kinase, SREBP1: sterol regulatory element-binding protein 1, PPAR γ: peroxisome proliferatory-activated receptor γ, GLUT1: glucose transporter 1, FIP200: focal adhesion kinase family interacting protein of 200 kd.

mTOR's specific interaction with autophagic mechanisms has been clarified recently. mTOR interacts with the ULK complex, consisting of the ULK1, FIP200, and ATG13 [52]. mTOR phosphorylates ATG 13, inhibiting the function of the ULK complex [53]. There has also been another proposed mechanism of mTOR autophagy regulation that involves the beclin1 complex. mTOR may inhibit activating molecule in Beclin 1-regulated autophagy protein (AMBRA1), a component of the beclin1 complex [54]. AMBRA1 activity is thought to enhance the ULK1 complex kinase activity [54]. In this way, autophagy is regulated at several points by mTOR, in response to cellular energy demands.

As stated earlier, autophagy is induced by cellular stressors, such as starvation. In a low nutrient or hypoxic environment, mTOR is inactivated, leading to the induction of autophagy [51]. In addition to inhibiting autophagy, mTOR leads to the activation of many metabolic processes, such as glucose metabolism and protein and lipid synthesis [6]. Namely, mTORC1 and mTORC2 increase glycolysis and increase glucose transporter 1 (GLUT1) expression [55]. mTOR is also thought to regulate lipogenesis, via regulation of the activity of peroxisome proliferatory-activated receptor (PPAR) γ [56].

6. mTOR in TB

mTOR's activity can be modulated by *Mtb* infection [57]. *Mtb* increases mTOR activity as measured by increased activity in downstream mTOR targets [57]. This is thought to be why there is an increase in cellular aerobic glycolysis [57]. This increase in glucose metabolism is also thought to be a key step in mounting a sufficient immune response against *Mtb* [57]. This metabolic shift during a *Mtb* infection is similar to that of the Warburg effect in cancer cells [58]. The Warburg effect occurs when cancer cells preferentially metabolize glucose by glycolysis, producing lactate, despite having adequate oxygen to undergo oxidative phosphorylation [59].

Another defense mechanism that could be effective in fighting an *Mtb* infection, is the induction of autophagy in granulomas. As discussed earlier, granulomas are a key process in walling off and fighting an *Mtb* infection, due to their ability to foster a beneficial environment for immune cells, as well as provide a physical barrier that prevents the infection from spreading. Additionally, however, granulomas may also provide an environment that fosters autophagic mechanisms. The hypoxic environment in specific types of granulomas has been shown to inhibit mTOR, inducing autophagy, as measured by increased levels of key autophagic enzymes [60].

7. Treatment

Several promising HDT strategies exist for the fight against *Mtb*. Strategies that specifically enhance autophagy can be divided into two categories—those involved in inhibiting the mTOR pathway and those that are not. Several points in the AKT/mTOR pathway might be targeted to promote autophagy, such as mTOR itself or AMPK. Direct inhibition of the mTOR complex by rapamycin and its analogs, also referred to as "rapalogs," is a well-established mechanism to promote autophagy [61]. Activation of AMPK by Metformin also promotes autophagy, via inhibition of mTORC1 (although its role in *Mtb* therapy needs further research) [62]. Drugs that inhibit ATGs and PIK3C3 are also being developed, although inhibition of these enzymes may not completely stop autophagy from occurring [4].

There are also several mTOR-independent targets which can promote autophagy. Ca+ channel blockers, such as Clonidine and Minoxidil, have been found to induce autophagy by increasing the levels of LC3 [63]. Several antipsychotics, such as lithium and valproic acid, also act to increase autophagy by decreasing levels of myo-inositol-1,4,5-triphosphate (IP3), which is thought to promote autophagy, although this mechanism is poorly understood [64]. Lithium has been shown to inhibit the growth of other mycobacterium species, but further studies need to be performed to establish its role in the pathogenesis of *Mtb* [65]. Numerous other drugs have proposed autophagy inducing mechanisms, such as resveratrol, spermidine, EGFR antagonists, vitamin D, and drugs that affect Beclin1 or nitrous oxide (NO) [3,4], although more investigation needs to be done on their therapeutic effect, specifically for *Mtb*. Further discussion of these drugs is outside of the scope of this review, but it is worth mentioning the numerous potential targets of HDT in the search of better, more effective treatments of *Mtb*.

This review will focus specifically on mTOR inhibitors as a potential therapeutic for *Mtb* treatment. Inhibition of mTOR by rapamycin analogs promotes autophagy, increasing the macrophages' ability to fight Mtb infection. Rapamycin analogs include sirolimus, temsirolimus, and everolimus [61]. These drugs have traditionally been used as anti-cancer treatments, due to their growth-suppressing effects, however, they are being investigated for the treatment of TB, due to their effect on the AKT/mTOR pathway and their autophagy. Of these three drugs, everolimus presents as a good candidate for further investigation as therapy for *Mtb* infection. Everolimus is a novel inhibitor of mTOR that could potentially be used as a therapy for *Mtb* infection and has been shown to decrease *Mtb* growth [6]. Everolimus is administered as an oral tablet and has a lower side effect profile than its injectable counterpart temsirolimus [66]. Additionally, everolimus has a greater bioavailability than sirolimus and it decreases vascular inflammation, more so than sirolimus [67].

The beneficial effects of everolimus on autophagy must be carefully weighed against its effects on the immune system, when treating *Mtb*. At high doses, everolimus is an effective immunosuppressant and is FDA approved for organ transplant recipients and cancer patients, but at lower doses it has shown to have an augmentative effect on host immune response [68–70]. A study conducted in 2014 showed that a group of healthy elderly individuals, treated with everolimus, showed a 20% improvement in their protective response after an influenza vaccination. This response included the reduced expression of programmed cell death-1 receptor on CD8+ and CD4+ T-cells via the inhibition of mTOR [69]. The subjects in this study were treated at a lower dose of everolimus than the transplant patients who are conversely at an increased risk of *Mtb* infection when using everolimus [6]. A proposed mechanism for delivery of everolimus to target cells infected with *Mtb*, without causing systemic immunosuppression, is through an inhaled nanoparticle preparation [71]. An in vitro study of inhalable rapamycin showed a more effective intracellular clearing of mycobacterium than rapamycin in solution [72]. These findings suggest that rapalogs, such as everolimus, might have a promising future as an HDT against TB.

8. Conclusions

Mtb infections pose a major global public health threat. Current treatments still revolve around antibiotic DOTS therapy. As antibiotic resistance grows other therapeutic targets will become more and more essential. HDT is a novel treatment strategy, aimed at using host immune mechanisms to battle infection. In this paper, we investigated the current literature on the AKT/mTOR pathway and autophagy, and their role in the pathogenesis of *Mtb*. We also investigated the current literature on everolimus as a novel therapy for *Mtb* infection, modulating cellular autophagic mechanisms via the inhibition of mTOR. The benefits of everolimus include less dependence on the use of DOTS therapy and the growing threat of resistant *Mtb*. These benefits must be carefully weighed against the immunosuppressive effect of everolimus. Novel drug delivery systems, such as inhaled nanoparticles might address this, although understanding the risks of treatment with each individual patient must also be carefully considered. Continued investigation of these novel therapeutic targets is crucial to addressing the global threat of *Mtb*.

Author Contributions: All authors contributed equally to the writing, editing, and review of the literature necessary to compose this paper.

Acknowledgments: The authors of this paper would like to thank Vishwanath Venketaraman for his guidance in writing this manuscript.

Conflicts of Interest: The authors have no conflicts of interest to disclose.

References

1. Barberis, I.; Bragazzi, N.L.; Galluzzo, L.; Martini, M. The history of tuberculosis: From the first historical records to the isolation of Koch's bacillus. *J. Prev. Med. Hyg.* **2017**, *58*, E9–E12.
2. Mittal, C.; Gupta, S. Noncompliance to DOTS: How it can be Decreased. *Indian J. Community Med. Off. Publ. Indian Assoc. Prev. Soc. Med.* **2011**, *36*, 27–30. [CrossRef]
3. Kolloli, A.; Subbian, S. Host-Directed Therapeutic Strategies for Tuberculosis. *Front. Med.* **2017**, *4*, 171. [CrossRef]
4. Rubinsztein, D.C.; Codogno, P.; Levine, B. Autophagy modulation as a potential therapeutic target for diverse diseases. *Nat. Rev. Drug Discov.* **2012**, *11*, 709–730. [CrossRef]
5. Sharma, V.; Verma, S.; Seranova, E.; Sarkar, S.; Kumar, D. Selective Autophagy and Xenophagy in Infection and Disease. *Front. Cell Dev. Biol.* **2018**, *6*, 147. [CrossRef]
6. Singh, P.; Subbian, S. Harnessing the mTOR Pathway for Tuberculosis Treatment. *Front. Microbiol.* **2018**, *9*, 70. [CrossRef]
7. Levine, B.; Mizushima, N.; Virgin, H.W. Autophagy in immunity and inflammation. *Nature* **2011**, *469*, 323–335. [CrossRef]
8. Itakura, E.; Mizushima, N. Characterization of autophagosome formation site by a hierarchical analysis of mammalian Atg proteins. *Autophagy* **2010**, *6*, 764–776. [CrossRef]
9. Mizushima, N. The role of the Atg1/ULK1 complex in autophagy regulation. *Curr. Opin. Cell Biol.* **2010**, *22*, 132–139. [CrossRef]
10. Axe, E.L.; Walker, S.A.; Manifava, M.; Chandra, P.; Roderick, H.L.; Habermann, A.; Griffiths, G.; Ktistakis, N.T. Autophagosome formation from membrane compartments enriched in phosphatidylinositol 3-phosphate and dynamically connected to the endoplasmic reticulum. *J. Cell Biol.* **2008**, *182*, 685–701. [CrossRef]
11. Watson, R.O.; Manzanillo, P.S.; Cox, J.S. Extracellular *M. tuberculosis* DNA Targets Bacteria for Autophagy by Activating the Host DNA-Sensing Pathway. *Cell* **2012**, *150*, 803–815. [CrossRef]
12. Wong, K.-W. The Role of ESX-1 in *Mycobacterium tuberculosis* Pathogenesis. *Microbiol. Spectr.* **2017**, *5*. [CrossRef]
13. Paik, S.; Kim, J.K.; Chung, C.; Jo, E.-K. Autophagy: A new strategy for host-directed therapy of tuberculosis. *Virulence* **2018**, 1–12. [CrossRef]
14. Manzanillo, P.S.; Ayres, J.S.; Watson, R.O.; Collins, A.C.; Souza, G.; Rae, C.S.; Schneider, D.S.; Nakamura, K.; Shiloh, M.U.; Cox, J.S. The ubiquitin ligase parkin mediates resistance to intracellular pathogens. *Nature* **2013**, *501*, 512–516. [CrossRef]
15. Franco, L.H.; Nair, V.R.; Scharn, C.R.; Xavier, R.J.; Torrealba, J.R.; Shiloh, M.U.; Levine, B. The Ubiquitin Ligase Smurf1 Functions in Selective Autophagy of *Mycobacterium tuberculosis* and Anti-tuberculous Host Defense. *Cell Host Microbe* **2017**, *21*, 59–72. [CrossRef]
16. Chauhan, S.; Kumar, S.; Jain, A.; Ponpuak, M.; Mudd, M.H.; Kimura, T.; Choi, S.W.; Peters, R.; Mandell, M.; Bruun, J.-A.; et al. TRIMs and Galectins Globally Cooperate and TRIM16 and Galectin-3 Co-direct Autophagy in Endomembrane Damage Homeostasis. *Dev. Cell* **2016**, *39*, 13–27. [CrossRef]
17. Chauhan, S.; Mandell, M.A.; Deretic, V. IRGM governs the core autophagy machinery to conduct antimicrobial defense. *Mol. Cell* **2015**, *58*, 507–521. [CrossRef]
18. Galluzzi, L.; Baehrecke, E.H.; Ballabio, A.; Boya, P.; Bravo-San Pedro, J.M.; Cecconi, F.; Choi, A.M.; Chu, C.T.; Codogno, P.; Colombo, M.I.; et al. Molecular definitions of autophagy and related processes. *EMBO J.* **2017**, *36*, 1811–1836. [CrossRef]
19. Sanjuan, M.A.; Dillon, C.P.; Tait, S.W.G.; Moshiach, S.; Dorsey, F.; Connell, S.; Komatsu, M.; Tanaka, K.; Cleveland, J.L.; Withoff, S.; et al. Toll-like receptor signalling in macrophages links the autophagy pathway to phagocytosis. *Nature* **2007**, *450*, 1253–1257. [CrossRef]
20. Martinez, J.; Malireddi, R.S.; Lu, Q.; Cunha, L.D.; Pelletier, S.; Gingras, S.; Orchard, R.; Guan, J.-L.; Tan, H.; Peng, J.; et al. Molecular characterization of LC3-associated phagocytosis (LAP) reveals distinct roles for Rubicon, NOX2, and autophagy proteins. *Nat. Cell Biol.* **2015**, *17*, 893–906. [CrossRef]
21. Bandyopadhyay, U.; Overholtzer, M. LAP: The protector against autoimmunity. *Cell Res.* **2016**, *26*, 865–866. [CrossRef]
22. Saleh, M.; Longhi, G. Macrophage Infection by Mycobacteria. *Mycobact. Dis.* **2016**, *6*. [CrossRef]

23. Reiner, N.E. Altered cell signaling and mononuclear phagocyte deactivation during intracellular infection. *Immunol. Today* **1994**, *15*, 374–381. [CrossRef]
24. Vergne, I.; Fratti, R.A.; Hill, P.J.; Chua, J.; Belisle, J.; Deretic, V. *Mycobacterium tuberculosis* Phagosome Maturation Arrest: Mycobacterial Phosphatidylinositol Analog Phosphatidylinositol Mannoside Stimulates Early Endosomal Fusion. *Mol. Biol. Cell* **2004**, *15*, 751–760. [CrossRef]
25. Deretic, V.; Singh, S.; Master, S.; Harris, J.; Roberts, E.; Kyei, G.; Davis, A.; Haro, S.D.; Naylor, J.; Lee, H.-H.; et al. *Mycobacterium tuberculosis* inhibition of phagolysosome biogenesis and autophagy as a host defence mechanism. *Cell. Microbiol.* **2006**, *8*, 719–727. [CrossRef]
26. Vandal, O.H.; Nathan, C.F.; Ehrt, S. Acid Resistance in *Mycobacterium tuberculosis*. *J. Bacteriol.* **2009**, *191*, 4714–4721. [CrossRef]
27. Deretic, V. Autophagy, an immunologic magic bullet: *Mycobacterium tuberculosis* phagosome maturation block and how to bypass it. *Future Microbiol.* **2008**, *3*, 517–524. [CrossRef]
28. Lopez-Castejon, G.; Brough, D. Understanding the mechanism of IL-1β secretion. *Cytokine Growth Factor Rev.* **2011**, *22*, 189–195. [CrossRef]
29. Chizzolini, C.; Chicheportiche, R.; Burger, D.; Dayer, J.M. Human Th1 cells preferentially induce interleukin (IL)-1beta while Th2 cells induce IL-1 receptor antagonist production upon cell/cell contact with monocytes. *Eur. J. Immunol.* **1997**, *27*, 171–177. [CrossRef]
30. Gutierrez, M.G.; Master, S.S.; Singh, S.B.; Taylor, G.A.; Colombo, M.I.; Deretic, V. Autophagy Is a Defense Mechanism Inhibiting BCG and *Mycobacterium tuberculosis* Survival in Infected Macrophages. *Cell* **2004**, *119*, 753–766. [CrossRef]
31. Taylor, G.A.; Feng, C.G.; Sher, A. p47 GTPases: Regulators of immunity to intracellular pathogens. *Nat. Rev. Immunol.* **2004**, *4*, 100–109. [CrossRef]
32. Harris, J. Autophagy and cytokines. *Cytokine* **2011**, *56*, 140–144. [CrossRef]
33. Caruso, A.M.; Serbina, N.; Klein, E.; Triebold, K.; Bloom, B.R.; Flynn, J.L. Mice Deficient in CD4 T Cells Have Only Transiently Diminished Levels of IFN-γ, Yet Succumb to Tuberculosis. *J. Immunol.* **1999**, *162*, 5407–5416.
34. Boom, W.H.; Canaday, D.H.; Fulton, S.A.; Gehring, A.J.; Rojas, R.E.; Torres, M. Human immunity to *M. tuberculosis*: T cell subsets and antigen processing. *Tuberculosis* **2003**, *83*, 98–106. [CrossRef]
35. Cooper, A.M.; Magram, J.; Ferrante, J.; Orme, I.M. Interleukin 12 (IL-12) Is Crucial to the Development of Protective Immunity in Mice Intravenously Infected with *Mycobacterium tuberculosis*. *J. Exp. Med.* **1997**, *186*, 39–45. [CrossRef]
36. Cooper, A.M.; Dalton, D.K.; Stewart, T.A.; Griffin, J.P.; Russell, D.G.; Orme, I.M. Disseminated tuberculosis in interferon gamma gene-disrupted mice. *J. Exp. Med.* **1993**, *178*, 2243–2247. [CrossRef]
37. North, R.J.; Jung, Y.-J. Immunity to Tuberculosis. *Annu. Rev. Immunol.* **2004**, *22*, 599–623. [CrossRef]
38. O'Garra, A.; Britton, W.J. Cytokines in Tuberculosis. In *Handbook of Tuberculosis*; John Wiley & Sons, Ltd: Hoboken, NJ, USA, 2017; pp. 185–225.
39. Algood, H.M.S.; Chan, J.; Flynn, J.L. Chemokines and tuberculosis. *Cytokine Growth Factor Rev.* **2003**, *14*, 467–477. [CrossRef]
40. Flynn, J.L.; Goldstein, M.M.; Chan, J.; Triebold, K.J.; Pfeffer, K.; Lowenstein, C.J.; Schrelber, R.; Mak, T.W.; Bloom, B.R. Tumor necrosis factor-α is required in the protective immune response against *Mycobacterium tuberculosis* in mice. *Immunity* **1995**, *2*, 561–572. [CrossRef]
41. Bean, A.G.D.; Roach, D.R.; Briscoe, H.; France, M.P.; Korner, H.; Sedgwick, J.D.; Britton, W.J. Structural Deficiencies in Granuloma Formation in TNF Gene-Targeted Mice Underlie the Heightened Susceptibility to Aerosol *Mycobacterium tuberculosis* Infection, Which Is Not Compensated for by Lymphotoxin. *J. Immunol.* **1999**, *162*, 3504–3511.
42. Valdivia, A.; Ly, J.; Gonzalez, L.; Hussain, P.; Saing, T.; Islamoglu, H.; Pearce, D.; Ochoa, C.; Venketaraman, V. Restoring Cytokine Balance in HIV-Positive Individuals with Low CD4 T Cell Counts. *AIDS Res. Hum. Retroviruses* **2017**, *33*, 905–918. [CrossRef]
43. Lagman, M.; Ly, J.; Saing, T.; Kaur Singh, M.; Vera Tudela, E.; Morris, D.; Chi, P.-T.; Ochoa, C.; Sathananthan, A.; Venketaraman, V. Investigating the causes for decreased levels of glutathione in individuals with type II diabetes. *PLoS ONE* **2015**, *10*, e0118436. [CrossRef]
44. Teskey, G.; Cao, R.; Islamoglu, H.; Medina, A.; Prasad, C.; Prasad, R.; Sathananthan, A.; Fraix, M.; Subbian, S.; Zhong, L.; et al. The Synergistic Effects of the Glutathione Precursor, NAC and First-Line Antibiotics in the Granulomatous Response against *Mycobacterium tuberculosis*. *Front. Immunol.* **2018**, *9*, 2069. [CrossRef]

45. Teskey, G.; Abrahem, R.; Cao, R.; Gyurjian, K.; Islamoglu, H.; Lucero, M.; Martinez, A.; Paredes, E.; Salaiz, O.; Robinson, B.; et al. Chapter Five—Glutathione as a Marker for Human Disease. In *Advances in Clinical Chemistry*; Makowski, G S., Ed.; Elsevier: Amsterdam, The Netherlands, 2018; Volume 87, pp. 141–159.
46. Crotzer, V.L.; Blum, J.S. Autophagy and its role in MHC-mediated antigen presentation. *J. Immunol. Baltim. Md 1950* **2009**, *182*, 3335–3341. [CrossRef]
47. Dörfel, D.; Appel, S.; Grünebach, F.; Weck, M.M.; Müller, M.R.; Heine, A.; Brossart, P. Processing and presentation of HLA class I and II epitopes by dendritic cells after transfection with in vitro-transcribed MUC1 RNA. *Blood* **2005**, *105*, 3199–3205. [CrossRef]
48. Brazil, M.I.; Weiss, S.; Stockinger, B. Excessive degradation of intracellular protein in macrophages prevents presentation in the context of major histocompatibility complex class II molecules. *Eur. J. Immunol.* **1997**, *27*, 1506–1514. [CrossRef]
49. Jagannath, C.; Lindsey, D.R.; Dhandayuthapani, S.; Xu, Y. Hunter, R.L.; Eissa, N.T. Autophagy enhances the efficacy of BCG vaccine by increasing peptide presentation in mouse dendritic cells. *Nat. Med.* **2009**, *15*, 267–276. [CrossRef]
50. Kim, Y.C.; Guan, K.-L. mTOR: A pharmacologic target for autophagy regulation. *J. Clin. Investig.* **2015**, *125*, 25–32. [CrossRef]
51. Jung, C.H.; Ro, S.-H.; Cao, J.; Otto, N.M.; Kim, D.-H. mTOR regulation of autophagy. *FEBS Lett.* **2010**, *584*, 1287–1295. [CrossRef]
52. Ganley, I.G.; Lam, D.H.; Wang, J.; Ding, X.; Chen, S.; Jiang, X. ULK1·ATG13·FIP200 Complex Mediates mTOR Signaling and Is Essential for Autophagy. *J. Biol. Chem.* **2009**, *284*, 12297–12305. [CrossRef]
53. Jung, C.H.; Jun, C.B.; Ro, S.-H.; Kim, Y.-M.; Otto, N.M.; Cao, J.; Kundu, M.; Kim, D.-H. ULK-Atg13-FIP200 Complexes Mediate mTOR Signaling to the Autophagy Machinery. *Mol. Biol. Cell* **2009**, *20*, 1992–2003. [CrossRef]
54. Nazio, F.; Strappazzon, F.; Antonioli, M.; Bielli, P.; Cianfanelli, V.; Bordi, M.; Gretzmeier, C.; Dengjel, J.; Piacentini, M.; Fimia, G.M.; et al. mTOR inhibits autophagy by controlling ULK1 ubiquitylation, self-association and function through AMBRA1 and TRAF6. *Nat. Cell Biol.* **2013**, *15*, 406–416. [CrossRef]
55. Zeng, H.; Cohen, S.; Guy, C.; Shrestha, S.; Neale, G.; Brown, S.A.; Cloer, C.; Kishton, R.J.; Gao, X.; Youngblood, B.; et al. mTORC1 and mTORC2 Kinase Signaling and Glucose Metabolism Drive Follicular Helper T Cell Differentiation. *Immunity* **2016**, *45*, 540–554. [CrossRef]
56. Kim, J.E.; Chen, J. Regulation of Peroxisome Proliferator–Activated Receptor-γ Activity by Mammalian Target of Rapamycin and Amino Acids in Adipogenesis. *Diabetes* **2004**, *53*, 2748–2756. [CrossRef]
57. Lachmandas, E.; Beigier-Bompadre, M.; Cheng, S.; Kumar, V.; van Laarhoven, A.; Wang, X.; Ammerdorffer, A.; Boutens, L.; de Jong, D.; Kanneganti, T.; et al. Rewiring cellular metabolism via the AKT/mTOR pathway contributes to host defense against *Mycobacterium tuberculosis* in human and murine cells. *Eur. J. Immunol.* **2016**, *46*, 2574–2586. [CrossRef]
58. Venketaraman, V. *Understanding the Host Immune Response Against Mycobacterium Tuberculosis Infection*; Springer: Berlin, Germany, 2018.
59. Courtnay, R.; Ngo, D.C.; Malik, N.; Ververis, K.; Tortorella, S.M.; Karagiannis, T.C. Cancer metabolism and the Warburg effect: The role of HIF-1 and PI3K. *Mol. Biol. Rep.* **2015**, *42*, 841–851. [CrossRef]
60. Huang, H.Y.; Wang, W.C.; Lin, P.Y.; Huang, C.P.; Chen, C.Y.; Chen, Y.K. The roles of autophagy and hypoxia in human inflammatory periapical lesions. *Int. Endod. J.* **2018**, *51*, e125–e145. [CrossRef]
61. Liu, Q.; Thoreen, C.; Wang, J.; Sabatini, D.; Gray, N.S. mTOR Mediated Anti-Cancer Drug Discovery. *Drug Discov. Today Ther. Strateg.* **2009**, *6*, 47–55. [CrossRef]
62. Restrepo, B.I. Metformin: Candidate host-directed therapy for tuberculosis in diabetes and non-diabetes patients. *Tuberculosis* **2016**, *101*, S69–S72. [CrossRef]
63. Williams, A.; Sarkar, S.; Cuddon, P.; Ttofi, E.K.; Saiki, S.; Siddiqi, F.H.; Jahreiss, L.; Fleming, A.; Pask, D.; Goldsmith, P.; et al. Novel targets for Huntington's disease in an mTOR-independent autophagy pathway. *Nat. Chem. Biol.* **2008**, *4*, 295–305. [CrossRef]
64. Sarkar, S.; Floto, R.A.; Berger, Z.; Imarisio, S.; Cordenier, A.; Pasco, M.; Cook, L.J.; Rubinsztein, D.C. Lithium induces autophagy by inhibiting inositol monophosphatase. *J. Cell Biol.* **2005**, *170*, 1101–1111. [CrossRef]
65. Sohn, H.; Kim, K.; Lee, K.-S.; Choi, H.-G.; Lee, K.-I.; Shin, A.-R.; Kim, J.-S.; Shin, S.J.; Song, C.-H.; Park, J.-K.; et al. Lithium inhibits growth of intracellular *Mycobacterium kansasii* through enhancement of macrophage apoptosis. *J. Microbiol.* **2014**, *52*, 299–306. [CrossRef]

66. Klümpen, H.-J.; Beijnen, J.H.; Gurney, H.; Schellens, J.H.M. Inhibitors of mTOR. *Oncologist* **2010**, *15*, 1262–1269. [CrossRef]
67. Klawitter, J.; Nashan, B.; Christians, U. Everolimus and Sirolimus in Transplantation-Related but Different. *Expert Opin. Drug Saf.* **2015**, *14*, 1055–1070. [CrossRef]
68. First Generic Drug Approvals. Available online: https://www.fda.gov/Drugs/DevelopmentApprovalProcess/HowDrugsareDevelopedandApproved/DrugandBiologicApprovalReports/ANDAGenericDrugApprovals/default.htm (accessed on 10 January 2019).
69. Mannick, J.B.; Del Giudice, G.; Lattanzi, M.; Valiante, N.M.; Praestgaard, J.; Huang, B.; Lonetto, M.A.; Maecker, H.T.; Kovarik, J.; Carson, S.; et al. mTOR inhibition improves immune function in the elderly. *Sci. Transl. Med.* **2014**, *6*, 268ra179. [CrossRef]
70. AFINITOR (Everolimus) Tablets [Package Insert] 2016. Available online: https://www.accessdata.fda.gov/drugsatfda_docs/label/2016/022334s036lbl.pdf (accessed on 13 January 2019).
71. Bento, C.F.; Empadinhas, N.; Mendes, V. Autophagy in the Fight against Tuberculosis. *DNA Cell Biol.* **2015**, *34*, 228–242. [CrossRef]
72. Gupta, A.; Pant, G.; Mitra, K.; Madan, J.; Chourasia, M.K.; Misra, A. Inhalable Particles Containing Rapamycin for Induction of Autophagy in Macrophages Infected with *Mycobacterium tuberculosis*. *Mol. Pharm.* **2014**, *11*, 1201–1207. [CrossRef]

MDPI
St. Alban-Anlage 66
4052 Basel
Switzerland
Tel. +41 61 683 77 34
Fax +41 61 302 89 18
www.mdpi.com

Journal of Clinical Medicine Editorial Office
E-mail: jcm@mdpi.com
www.mdpi.com/journal/jcm

www.ingramcontent.com/pod-product-compliance
Lightning Source LLC
LaVergne TN
LVHW070044120526
838202LV00101B/428